The Speed Trap

ALSO BY JOSEPH BAILEY

The Serenity Principle
Finding Inner Peace in Recovery
1990, HarperCollins

Slowing Down to the Speed of Life
How to Create a More Peaceful, Simpler Life from the Inside Out
Co-authored by Richard Carlson, Ph.D.
1997, HarperSanFrancisco

The Speed Trap

How to Avoid the Frenzy of the Fast Lane

Joseph Bailey

 HarperSanFrancisco
A Division of HarperCollins*Publishers*

HarperCollins books may be purchased for educational, business, or sales promotional use. For information please write: Special Markets Department, HarperCollins Publishers, Inc., 10 East 53rd Street, New York, NY 10022.

HarperCollins Web Site: http://www.harpercollins.com

FIRST EDITION
Designed by Joseph Rutt

Library of Congress Cataloging-in-Publication Data
Bailey, Joseph V.
The speed trap : how to avoid the frenzy of the fast lane / Joseph Bailey. — 1st ed.
ISBN 0-06-251589-6 (pbk.)
1. Conduct of life. 2. Quality of life. 3. Conduct of life—Case studies. 4. Quality of life—Case studies. I. Title.
BF637.C5B34 1999
158—dc21 98-47952

99 00 01 02 03 ❖/HADD 10 9 8 7 6 5 4 3 2 1

To my family, whose love has made slowing down to the moment in my own life essential and a true joy. To Michael, my wife. To Ben and Michelle, my son and daughter-in-law. To my soon-to-be-born grandchild. And to my mom—my biggest fan and source of support. You are all a gift!

CONTENTS

ACKNOWLEDGMENTS

This book would not have been possible without the joint efforts of many people. First, I want to thank all the people who agreed to be interviewed for this book. Although their names have been changed to preserve anonymity, their lives and personal transformations will inspire many people throughout the world. Second, I want to thank my editor at Harper San Francisco, Mark Chimsky, who asked me to write this book. He has been an ideal editor—challenging, encouraging, and helpful during each step of the process. Third, I want to thank Laurie Viera, who edited my manuscript before my publisher ever saw it. She was gifted in her ability to magically transform my words into prose and was a pleasure to work with. Fourth, I want to thank my dear friend and colleague Richard Carlson for encouraging me and believing in me as a writer, and for writing the foreword for this book. Last, I want to thank my family for all their loving support, especially Michael, my wife, who is my best friend and soul mate in the truest sense of the word.

FOREWORD BY RICHARD CARLSON

My family and I were vacationing in Hawaii, one of the most beautiful and relaxing places on earth, when I overheard a mother yelling to her two children in an impatient tone, "Hurry up. You're taking too long." They were headed toward the beach to play in the waves! I've witnessed businessmen and businesswomen start and finish entire conversations so preoccupied and rushed that they never, even once, looked up to verify whom they were speaking with. And I've seen restaurant customers so angry at having to wait a few minutes for a table that they stormed out, cursing and vowing never to return. It seems that impatience, as well as a corresponding sense of hurry, has become an epidemic. Indeed, the most popular words and phrases in the English language now seem to include "Hurry up, I'm late," "There's not enough time," and, everyone's favorite, "I'm really busy."

This being the case, I can't think of a more appropriate or important book for our times than this one. In my view, addressing the issue of our obsession with speed, our tendency to glorify it, and the way in which we become entrapped by it is critical. As with other aspects of life, there is a common misconception regarding speed that "More is better" or, more specifically, "Faster is better." We believe that moving quickly, traveling quickly, communicating quickly, eating and thinking quickly, even having fun quickly is the way to go.

Our obsession with speed has taken over our lives and, without question, has diminished our quality of life. Many of us have become too busy and hurried to relax, to listen to others, or even to appreciate friends and loved ones. We are often so anxious to get to "what's next" that we fail to fully experience or appreciate what we are already doing.

One of my favorite quotes says, "Progress was great once, but we went too far." Not always, but often this is true with regard to speed. Certainly there are times when faster *is* better. Speedier air travel and many technological advances have surely enhanced life and made it a lot easier and more convenient. Yet we've taken the notion of faster performance to a level, almost, of insanity.

Our obsession with speed has caused us to become increasingly impatient and agitated and has created a societal lack of perspective. We insist on instant gratification. If our computer takes more than a few seconds to boot up, we're annoyed. If our favorite fast-food establishment takes more than a minute to prepare our meal, we impatiently look at our watch, wondering what's wrong. We ask ourselves, "Why is this taking so long?" If a friend is a few minutes late, we become bothered—we might even wonder why we "waste our time" with friends like that! Gratitude is a thing of the past. We're too busy and hurried to stop and smell the roses, to slow down long enough to notice many of life's simple pleasures—nature, sunsets, friends and family, health, our many blessings—which we take so much for granted.

Our minds have become so incredibly busy that we don't know how to relax. Instead, we become easily bored when something exciting or "interesting" isn't happening. Many of us need constant stimulus. I travel on airplanes quite a bit. I often see fellow passengers reading not one, but two newspapers at the same time—while they watch the movie and, sometimes, even talk on the phone. Almost no one is able to sit still without being entertained, even for a moment. We've become frightened of, and uncomfortable with, silence. Our minds are buzzing so loudly that, sometimes, it almost seems as if I can hear the confusion and anxiety in the mind of the person sitting beside me! I often observe others twitching nervously and emitting frequent sighs of impatience.

Sadly, our speeded-up minds and our superspeeded-up lives have profoundly negative effects on families. We no longer have time for each other. We rationalize by saying we are spending "quality time," but in reality "quality time" is usually nothing of the sort. Rather, it is often nothing more than "scheduled time," a brief moment between two or more speeded-up activities.

Intimate conversations are rare these days. For the most part, we've become "too busy" to sit still and enjoy the company of a friend, our spouse, even our children. Rather than looking at one another and paying close attention when we speak, one eye is usually on our watch, making sure we leave plenty of time to get to our next appointment. We fear we're going to fall behind or miss something important. The irony is, the faster you go, the more you will miss!

Of course, we rationalize all of this speed by insisting we have become more productive. Again, not true. Speed, more than most other factors, causes mistakes, a lack of focus, overlaps in effort, and unnecessary problems that would never occur if we had simply slowed down enough to focus with a clear, uncluttered mind.

Although some authors recognize the trap associated with too much speed, the well-intended solutions often compound or exacerbate the problem, rather than fix it. Typically, we turn to time-saving gadgets, technology, or better ways to organize our lives. When we do manage to carve out some free time or become more efficient, thus saving time, our nervous, speeded-up minds always find a way to fill it. Therefore, we end up going in circles—trying hard to save time and then filling up the saved time with more activities.

As far as I'm concerned, the only legitimate solution is to create a calmer mind. This is the emphasis of *The Speed Trap*. In it, you will see how everyday life appears with a calmer outlook. A calmer mind brings with it freshness, insight, creativity, and the energy necessary to work long hours if that is what's required. A calmer mind allows you to see right to the heart of the matter, to peel away what is irrelevant. A calm mind allows you to have a sharp and effective learning curve, to focus, concentrate, and to be your best. It allows you to connect with others on a deeper level, to develop a powerful presence and genuine rapport that brings forth nondefensive, effective behavior from others.

This book will be very comforting to you. You won't have to change your life or purchase a new time-management system. All the changes will occur in the peace and quiet of your own mind, through shifts in your awareness. As you absorb the material, you may find yourself asking questions like "What's the hurry?" All the rushing around may begin to seem a little silly. You'll discover that, in reality, there's plenty of time to get everything done. In fact, when you learn to slow down "from the inside out," it will seem as if you have *more* than adequate time, as if life is coming at you at a slower pace. You will learn to keep your bearings. You will become more efficient and less reactive, and you will make fewer mistakes. You will also become more patient, kind, and willing to listen.

Who better to share his insights on this important topic than Joseph Bailey? In the field of self-improvement, there is no better measure of a teacher's authenticity than the way that person lives his or her life. Sadly, there are authors, speakers, and teachers whose private lives are about as different from what they are teaching as you can imagine. Some who write about inner peace are anxious, angry, and agitated a great deal of the time. Some whose primary teaching or message revolves around kindness, charity, and acts of love are greedy, narcissistic, self-centered, and selfish. And some so-called relationship experts have been divorced not once or twice but many times, and seem to struggle in virtually all their relationships. Who knows? Perhaps this does make them experts on what not to do.

None of this dichotomy applies to Joseph Bailey. I've had the great honor of knowing Joe for a number of years, and we have become extremely close friends. As far as I'm concerned, if anyone is qualified to discuss this topic, he is. One of the things that almost always happens when Joe and I get together is that I realize how hyper and speeded up I can get at times. Although Joe is an extremely gifted therapist and an accomplished professional, he is one of the most relaxed people I've ever known. He's calm and reassuring, both in everyday life as well as under pressure. He's extremely "present" and kind to everyone he meets, and not just when people are watching, but all the time. He's as gentle, genuine, and polite to a homeless person as he is to the CEO of a giant corporation—I've seen him with both.

Joe's sense of calm extends to his marriage. He and his wife have a genuine, loving relationship that is nourishing and fun. Their mutual respect and tenderness for one another is obvious to all who know them. There is no question that all of us can learn a great deal from Joe Bailey.

Unfortunately, even Joe Bailey can't reduce the number of responsibilities you have or the demands of your hectic schedule. He can, however, help you discover a place within yourself where life doesn't seem so rushed or as much like an emergency. Gandhi once said, "Your life is your message." Indeed, Joe is a living example of what he teaches. I sincerely hope you'll take his message to heart.

My suggestion is this: Take a deep breath and prepare to relax. I love this book and know you will too!

Richard Carlson
September 1998

INTRODUCTION:

PEACE OF MIND IN A FRENZIED WORLD—PIPE DREAM OR POSSIBILITY?

As we approach the year 2000, speed has become our god. We worship efficiency, getting the job done fast, accomplishing more and more, beating out the competition, getting to the bottom of our to-do list, and multi-tasking to get it all done. Advertisers sell more products now with speed than they used to with sex. Now it's instant breakfast food, microwave waffles, instant cash and credit, and loans over the phone in the blink of an eye. We double the speed of our computers at least every two years, so we can get more done, keep up with the latest facts through the Internet, and send e-mail instantaneously to all our co-workers all over the globe NOW! We've got to keep our bodies moving faster, too. So we drink more espresso for a quick jolt, eat super vitamins to perk us up, and find the latest fad diet to have more energy. We might be getting more done than ever before, but are we enjoying our lives?

As a psychotherapist and seminar leader who has worked with thousands of people, the major complaints I hear from my patients are stress, anxiety, insomnia, relationship problems, and depression. All too often, my patients ask, "What's the point of all this rush to accomplish, acquire, and produce?"; "Why am I not happy?"; "I have everything I could ever want. I go on dream vacations, and I'm successful. My life is full, but I'm not fulfilled." I hear this litany of complaints every day in my office and on talk shows, yet we continue to plunge into the next century as though we are all in a race.

Is this really necessary? Is there another way?

The God of Speed

Twenty years ago, I lived my life on a treadmill. The faster I ran, the faster the speed of my treadmill. Each year of my life I tried harder and seemed to be busier, yet I felt as if there was always more to do. Speed had become my god. Inside I felt like a revved-up engine—anxious, stressed, worried whether or not my to-do list was done or would ever get done. I would feel satisfied and relieved for only an instant after accomplishing one more thing—a phone call returned, a day of clients seen, a talk given, one more load of laundry—but almost as instantly, my anxiety would begin to build up again. The result: I was living in fear and tension, and I suffered from insomnia and headaches. I was always in a hurry, I felt guilty about not getting enough done, and I basically felt dissatisfied with my life. I longed for weekends, vacations—or some elusive day when I could save up enough to retire and get out of the rat race. Sound familiar?

Then, about 1980, I got a call from an old graduate school friend who asked me to attend a seminar on a revolutionary new psychology called Health Realization,[1] which was based on the principles that governed mental health. I was already cynical of anything that professed to be a new paradigm and promised to have all the answers. I'd heard it all before many times and become disillusioned. But for some reason, I felt drawn to attend the seminar, almost compelled by my curiosity.

What I discovered that weekend has continued to unfold within me since that day. What I learned has allowed me to live my life without the stress and anxiety that had become "normal" to me over the course of my adult life. Despite what you might expect, I didn't have to drop out of society, quit my job, move to the north woods, and become a reclusive hermit. Not only did I stay focused on my career, I went from burnout to total rejuvenation. I have been able to accomplish things in my life that I never even dreamed were possible. I have success without stress, I have the love of my life without "working on" a relationship, and I have an inner contentment in my everyday life that I thought was reserved for monks in a monastery, retirees, and young children—not responsible adults in a fast-paced working society.

An Inside-Out World

What I discovered in that seminar is that my personal world, the world of my psychological experience of life, is created from the inside out. What I know now is that *every breathing moment I am creating my life psychologically through thinking*. Moment to moment, I am thinking

---

1 Health Realization has also been referred to as Psychology of Mind.

one thought after another—whether I know it or not—and each of those thoughts creates the moment-to-moment life I perceive, the emotions I feel, and the reactions I have to that perceived reality. This knowledge has given me a feeling of awesome responsibility and freedom, for I have discovered the gift of the *free will.*

I used to believe that life came to me from the *outside in.* I had been trained by all the influences of my life (school, parents, graduate psychology training) that my experience of life was a product of my upbringing, my past, and outside influences (society, the environment, the economy, politics, the media). And my experience confirmed this training. To me, it appeared that the weather, traffic, other people's moods and behavior, my checkbook balance, the stock market, and all other external forces conspired to dictate my inner experience. After all, when it (reality) changed, my inner experience changed, too. And so it seemed to me that there was a stationary, concrete, external, and objective reality to which all of us react.

I did realize that my thinking had something to do with my experience, but only that it was some kind of middleman between reality and me. *What I didn't realize is that the actual personal reality that I saw "out there" was totally a product of my power to think.* I believed that life "out there" was causing my stress; I was merely a victim of today's hectic world. I couldn't have been further from the truth.

A New Reality

When I first heard this notion that it was my thoughts—and my thoughts alone—that created my reality, I felt defensive and angry. After all, I had spent years learning how my family and society were the cause of my thinking and experience, and I was teaching these ideas to my psychotherapy clients, seminar participants, and other professionals. The new ideas sounded like simplistic and even dangerous platitudes. But, paradoxically, at some intuitive level, the ideas made absolute sense and gave me a feeling of deep calm and serenity. Obviously, I was totally confused.

Over the past twenty years since I first heard these ideas, they have withstood the test of time not only within my life, but also in the lives of thousands of people whom I have met, people exposed to those very same ideas. I have seen that as people realize the principles that I will explain in a moment, their lives become calmer, more fulfilling, wiser, and more loving—successful in every sense of the word. *By understanding that life is created from the inside out, we are empowered to create a beautiful life.* If we are here on this planet but a short time—maybe sixty, seventy, eighty years—why not benefit from this gift?

Why Speed Trap?

I will use the term speed trap throughout the book to communicate how we are all caught up to one degree or another in a busy mind—*a mind that is speeded up and out of the moment.* Whether you think you are going too fast and your life is unmanageable, or not—at the root of all our psychological problems lies a busy mind, trapped by its own thinking. The speed trap is the common denominator of humanity's problems.

How the Mind Works—Three Simple Principles

We are all part of the miracle we call life. As I look out the window and see the spring leaves budding again, the birds returning from their winter migration, and the daffodils in bloom, I witness the mystery of life at work—how life dies, transforms, and is reborn eternally. Whatever you choose to call this life-force that is behind all of nature, all the planets, stars, galaxies, and our own heartbeat, it is definitely at work all the time. Without that life-force, we could not experience all there is to experience, for we, too, are a part of that life-force. This life-force is called **Mind.**

The First Principle

Mind: The Energy of Life Itself—The Source of Thought and Consciousness.

Mind is the source of the invisible energy that we can't see, smell, or touch; it is behind everything. Without Mind, we wouldn't exist. It's behind all that is part of our psychological life as well—every thought, perception, emotion, action, movement, and intention. Mind itself has no form, but it is the source of all forms that come into existence—a smile, an invention, a great idea, a thoughtful gesture, a jealous rage. Mind is the power behind what we experience. We can never know Mind in its totality, but we can sense its presence in everything.

Think of Mind like electricity. We can't see it, yet we can see its effects in light, energy, movement of an engine, a bolt of lightning. Without the electricity, the engine wouldn't have any power or movement, the light bulb wouldn't shine. Mind is the source of all human experience—thoughts, perception, emotion, behavior, and awareness.

The Second Principle

Thought: The Continual Creation of Life through Mental Activity.

Thought is the principle that allows us to create the **form** *of our moment-to-moment experience of life.* Thought creates all mental activity—every mental image, fantasy, perception, sound, touch, pain, pleasure, emotion, sensation, concept, memory, dream, imagination, worry, all of what we think of as our experience. We are continually thinking, and each thought is continually giving us our moment-to-moment experience of life.

As I awake each morning, I might have a thought such as, "Oh, no, another Monday morning!" or "I wonder if my car will start" or "I can't wait to go outside and go for a run." The next instant I might think "God, this bed feels so good, I just love lying in bed . . ." or "I'm so lazy, why is it so hard to wake up?" or "Oh well, just get up, Joe, it won't kill you." With each passing thought, there is a change in the quality and intensity of my experience. Even my whole physiology changes with each thought, no matter how slightly. For example, when I have rushed and stressed thoughts, my brain sends electrical impulses to my stomach, shoulders, and jaw that lead to a feeling of physical tension, and thus I *feel* stressed.

Most of the thoughts that we think occur totally without our awareness. But whether or not we are aware of those thoughts, they will nonetheless become our experience. The thoughts we have instantly transform into what we call "reality." Most thoughts are not even what we would usually think of as thoughts. For example, noticing my back is tight from sitting in this chair in front of my computer and readjusting my posture is actually a series of thoughts. I notice the discomfort, then it occurs to me to sit up straighter or stretch, all without an awareness of the thoughts that created that experience. Or having the passing thought that I am running late for an important meeting might cause me to feel fear, step on the gas, and terminate my perception of the beauty of nature around me. I will only notice how slowly the traffic is moving or how poorly someone is driving in front of me. On the other hand, the person in the next lane in the same traffic is seeing all the spring buds bursting from the trees and is feeling elated to be alive. This is all thinking. Through the gift of thought we are little *reality generators*.

Separate Realities

Because we are always thinking and our thinking is continually generating new experiences, we are always creating a new reality. Furthermore, no two people are ever creating exactly the same reality, for no two people think in exactly the same way. Just as snowflakes are all made of the same substance, but no two are alike, we all create reality with our thinking, but we are always thinking differently. Therefore, each of us lives in a *unique personal reality*. None of us can really know another person's experience, because we can never have that person's exact same thoughts. Even two stage actors playing the same role with the same lines will each experience it differently. And each actor will even experience subtle differences in playing the role on different nights.

The Free Will

Whether you realize it or not, *thought is a choice*. You can change your thoughts at will, and even if you don't try to, thought is constantly changing nonetheless. Once you realize the principle of Thought at work in your life, you will see the power of the free will to change your experience of life.

For example, if I wake up and think, "Oh, another Monday morning!" and I realize I don't like the experience of where that thought is taking me, I can change it. However, I also have to realize that the *source* of my experience is not the fact that it is Monday, but that I am thinking that thought. *Simply realizing that I am the thinker of my thoughts allows my thoughts to change the instant I recognize that fact.* With this knowledge in mind, I can then say to myself, "This doesn't feel very good. I could ruin a whole day that way!" And in a instant, my mind will give me another thought.

To summarize, we create life with Thought, it is always at work, and it accounts for all of our experiences. No two people live in the same reality because we all have a unique thought in the moment. When you simply recognize the source of your experience—Thought—you are able to have a new thought. I'll talk more about this in a minute.

The Third Principle

Consciousness: The Continual Sensory Experience of Thought as Reality.

Consciousness is the principle that brings every thought to life through the senses. Without Consciousness, you would not experience your thinking as reality. Thought would be like unread words on a page of a movie script—meaningless and unexperienced. Consciousness is like a full sensory special effects department that creates a movie with great sound and visual effects, plus smells, kinesthetic sensations, emotions, the whole gamut of sensory experience. Virtual reality machines look amateurish when compared to the power of Consciousness to make thought appear as reality. As an example of consciousness in action, consider a father who thinks, "I am so lucky to be a parent. My child is so special." His heart warms, a smile appears on his face, his endorphin level rises, and he relaxes.

Consciousness is nondiscriminatory—it will bring any thought to life that passes through your mind. If I am walking down a dark path, begin to feel afraid of what might be lurking in the dark, and I see a stick on the ground, I might perceive the stick as a snake for an instant. While I am thinking that thought, I will have the sensory experience of a snake, and the experience will replicate my thought exactly (a rattlesnake will generate a stronger reaction than a harmless garden snake). If I imagine that my day is going to be hectic—in that instant I will experience a hectic feeling, even if I am still in bed. Perhaps I am in pain from a surgery, and someone whom I love dearly walks into the room. As long as

I fill my head with thoughts of love, my experience of pain will disappear during those moments—until that person asks me how I'm feeling. Then, I may think about pain, and consciousness will make the pain reappear. If I am rushed and running out the door to work, consciousness will bring rushed feelings to my body. If I have the thought the next moment that I am lucky to have a job, consciousness will bring the feeling of gratitude to my experience. *Consciousness is the breath of life of thought*.

If we think of our psychological experience of life like a movie, then *thought* is the script, the series of images on the film, and the sound on the soundtrack. Those words, sounds, and images are brought to life via our senses through *Consciousness*—the projector, speakers, and the theater—from the power generated through the electricity of *Mind*.

Feelings: Our Built-in Guidance System

If most of our thinking occurs without our awareness, how can we protect ourselves from harmful or unhealthy thinking? It's true that until you "wake up" and recognize that you are thinking and creating your moment-to-moment experience, there isn't much you can do about it. However, nature has built into our psychological apparatus a signal device to alert us when our thoughts are unhealthy, in the same way it provides us with pain sensors and discomfort when we are sick, hungry, tired, or have eaten something harmful.

When you are thinking in an unhealthy or harmful way such as stress, anxiety, rage, or resentment, your physical sensations will be thrown off balance, and you will experience some degree of discomfort. If you pay attention to these signals and simply notice that you are thinking, the thinking will change, adjust—and a new thought will come. Most people have been taught that feelings have a life of their own, but we now see that they are integrally connected to our thinking. Feelings are the physical manifestation of Thought + Consciousness.

If you learn to trust the warning signal of your feelings, they can be a sort of mental alarm clock that alerts you to wake up to your thinking. When you "wake up" from a series of harmful thoughts or even one harmful thought, your mind will clear, and a deeper intelligence will let you know whether or not to continue on that line of thought. We call this deeper intelligence *wisdom*.

Wisdom

Wisdom is thinking in the moment that is not bound by memory. It is a transcendent intelligence that seems to come from "out of the blue." It is thought that is not generated from a personal belief system, but rather from an impersonal source. Wisdom might appear in the form of an insight that comes into your mind about a problem that has been troubling you for a long time, but that you have been unable to resolve. Often when you have let go of the problem while sleeping, exercising, driving, in the shower, or in any other situation when your mind is disengaged from the problem, you will get an insight that will seem simple, obvious, and true.

For example, the other day I was doing a workshop with a group of university professors about stress and how they could be less frantic in their work and actually be more productive. After the first day, I had to plan day two, and I had asked them for input on what they would like to learn during the second day. When I returned to my motel room, I sat down and tried to come up with a plan. But I was so tired that it seemed painful to think about it. So I put it "on the back burner," rested, had dinner, and watched TV. Although I still had no creative insights about the next day, I went to sleep, trusting that my deeper, creative intelligence would deliver by morning. Sure enough, within ten minutes after I awoke I had a flood of creative ideas that were perfect for that day's training.

Over the years, I have learned to trust my inner wisdom through trial and lots of error. I have had many futile experiences of *trying to make* the creative thoughts come. But when I have had faith in my wisdom and let it go, that wisdom has always come through.

Some people seem to have wise thoughts much of the time, while others seem to have wise thoughts but rarely. It's not because some people are inherently wiser than others. It's because people seem to have more wise thoughts when they are relaxed, resting, recreating, having fun, and are in a positive state of mind. When our minds are relaxed and we feel content, we tend to have wiser thoughts. We have observed that when people can learn how to access this wisdom through faith in its existence and an ability to let go, they have less stress, more happiness, and greater success in professional and personal endeavors.

Innate Mental Well-Being

Where does wisdom come from? We are all born with innate mental well-being. Just observe a young child and you will witness mental well-being. It is the human capacity for the virtues of wisdom, self-esteem, creativity, unconditional love, motivation, adaptability to change, and the deeper feelings of joy, compassion, love, optimism, patience, and humor. These qualities come as a package deal. As with an acorn, the oak tree is complete within it, just not yet realized.

We have observed in our work with people of all socioeconomic groups, educational levels, mental illness diagnoses, and cultures, that once they gain an understanding of the way their mind works they can recover their innate mental health. In other words, we have found that innate mental well-being can be covered up but never destroyed. It is capable of resurfacing at any moment. Like a cork in the water, it rises when the weight of a speeded-up mind is removed. Our *experience* of our innate mental well-being fluctuates with the quality of our thinking.

The Quality of Thinking

Healthy thinking ←————————→ Unhealthy thinking

For all of us, our thinking seems to fluctuate on a continuum from healthy to unhealthy thinking. When our thinking is healthy, we tend to have insights, wisdom—thoughts that are positive, hopeful, new, and creative. When our thinking is unhealthy, we tend to have redundant thoughts that are circular and seem to lower our spirits. These thoughts tend to be more analytical and take us out of the moment, like worry, regret, and busy mindedness.

How can you tell if your thinking is healthy? When your thinking is healthy, you are in the moment. You are present in the here and now, and your thinking is responsive to that moment. If you are with your child, you are present, listening, and responsive to your child's needs. If you need to plan something in the future that needs to be done now, it will occur to you from a steady flow of thought. If you need to strategize, analyze, or plan, that will occur to you as well. And if you need to remember something from the past, that will also occur to you in the moment. In other words, your thoughts are flowing. You are in a state of *being*, and you are enjoying life at this very moment.

When you are in unhealthy thinking, you are not in the moment—you are in the past or the future. You live in a world of habit and memory, which is fixed, non-creative, and often unresponsive to what is happening in that moment. If you are with your child in a state of unhealthy thought, you might tend to be preoccupied, self-absorbed, impatient, angry, or judgmental. The child will feel this and will be clinging, defiant, or act out in some way. The quality of the relationship and the feelings will go down.

As discussed in the previous section, your feelings and sensations can allow you to recognize these fluctuations in the quality of your moment-to-moment thinking. As your thinking moves into an unhealthy quality, your feelings will tell you this through discomfort. If you see the source of your experience as coming from your thoughts, it will make sense to change your thinking.

However, you need not actively change your thinking. Thinking has a built-in default setting of wisdom.

The Leap of Faith

Trusting in your wisdom takes a leap of faith. There are no "techniques" to make this happen. Each of you confronts a moment of truth where you seem to have a choice, and you can then say, "I can keep worrying about this, or I can let it go and trust that once my mind is quiet, an answer or the thoughts I need will come to me." There are no guarantees that an answer will come. You just have to trust that it will, and see what happens. Each time you do this, your faith becomes stronger, and you know that wisdom is there, on your side, guiding you through life.

Over the years since I first heard these ideas, I had to discover for myself the truth of these principles. Each time it has gotten easier, to the point where now I have no doubt that the right thoughts will come. Knowing this gives me peace of mind. No one can take this step for you. Each of you must take the leap of faith and discover *for yourself* the power of thought and, in particular, the power of wisdom.

The Evolutionary Nature of Wisdom

Because each of us has this built-in flow of wisdom, we can learn what it feels like to relax as we make our way through life. As more and more wise thoughts come to you in the moment, they will tend to evolve your level of understanding of life. As your level of understanding goes up, you begin to take life less personally, you see reality as coming from the inside out, and you live with more faith. The reward for this is that your feelings will deepen and you will experience more joy, love, contentment, and peace than you ever imagined possible. There is no limit to the level of understanding available to us. Your life will continue to get better and better as your understanding deepens.

How to Read This Book

The purpose of this book is to help you raise your level of understanding of how life is created—how you as the thinker are the creator of your reality. As your understanding deepens, your mind will slow down, yet your life will become more manageable and successful. Remember, life comes from the inside out.

Rather than write a step-by-step self-help book or a theoretical book on the mind, I decided to write a book of stories about everyday life to illustrate these principles. On the pages that follow, there will be thirty-seven stories that convey the understanding I have presented in this introduction. Each story will have a principle, guideline, concept, or tip about how to break out of the frenzy of today's world. The stories will be about the kind of day-to-day challenges most of us face—getting caught in traffic, facing deadlines, getting the kids to school on time, dealing with a computer breakdown, and many other more serious life situations. These stories are based on real people, but the names have been changed to protect their anonymity.

You can read the stories in any order. **I do suggest you read this introductory chapter first and then come back to it from**

time to time to deepen your understanding of the principles. The principles are simple, but often challenging at first to grasp. As you read each story, the underlying principles will come to life for you. More importantly, as you go through your daily life, you will begin to see the principles at work in yourself and in others around you.

Read the stories as you would read a novel, rather than as you would a textbook. In other words, don't pressure yourself to "get it." When your mind is relaxed, the truth of the message will jump out from between the lines. Fables, fairy tales, and children's stories are often the most profound source of truth. They are enjoyable to listen to or read, but the deeper meaning often only sinks in later. This is a book of modern-day stories with which all of us can identify. They tell us how to find sanity and a calm mind in a hectic world.

The Promise of Breaking Out of the Speed Trap

If you take this message to heart, the following can happen for you, as it has for me and thousands of others who have gained this simple understanding:

- You will learn to slow down and enjoy each moment.
- You will learn to find calm inside, even in the midst of today's hectic world.
- You will begin to trust that your wisdom is guiding you through life, every step of the way.

- You will realize that you can let go of worry because your insight will protect you.
- You will discover that contrary to conventional wisdom, your productivity and efficiency will actually increase as your mind slows down to the moment.
- You will realize that other people, no matter how low their mood or how negative their behavior, don't have to ruin your day.
- You will know how to take the daily "crises" that happen to all of us in stride and handle them with common sense, instead of an emotional reaction.
- You will discover that you don't have to give up your lifestyle or drop out of society in order to have peace of mind.
- You will find that living in the moment is the best preparation for the unexpected.
- You will truly know from experience that you can be happy.

1

IS THE GRASS REALLY GREENER ON THE OTHER SIDE OF THE FENCE, OR DO I JUST NEED NEW GLASSES?

When we are unhappy, we often quite naturally begin to fantasize about how our troubles would end "if only." *If only I could move to the country and avoid the hassle of city life . . . If only I could quit my job and do something simpler, like be a forest ranger in a national park . . . If only I could leave this marriage, which isn't working out, and find someone who really understands me. . . .*

Unfortunately, all too often we quit that stressful job or leave that unhappy marriage, only to find ourselves in a similar or worse situation. Why is this?

Looking for external solutions to our psychological problems doesn't work. In other words, if we don't change our thinking, we will bring that thinking right along with us to the next job or the next marriage or the new house in the country. Our experience of life is the creation of our own thinking>perception>emotion>behavior. This doesn't mean that people should never change jobs or careers, move to a better location, or even find a new significant other. It simply means that nothing changes on the outside if nothing changes on the inside, where your experience is created—your mind.

Doug came to me for help with several problems. He was stressed beyond his limit. He was unable to sleep at night, hated the industry he was in and the people he worked with, had angry outbursts at work, and couldn't maintain a significant relationship. To Doug, it appeared that he was in the wrong job and living in the wrong state and that his problems were all a result of those factors. In the past ten years he'd taken five new jobs, moved to four new homes, and failed in numerous relationships.

Doug would often fantasize about moving to Colorado, getting a job at a ski resort, and simplifying his hectic life. He was making great money where he was, but the stress wasn't worth it to him, and he knew he wasn't getting any younger. He sought me out because he wanted to make sure he wasn't going to regret his decision later.

"Everybody around here is always pushing my buttons," Doug said. "They have no respect for my time, my priorities, or my responsibilities, yet they want me to respect theirs. Well, I tell you, they can take this job and do you know what with it!" This was Doug's typical complaint. But as he began to understand the principles described in this book, his view of his job and other people gradually began to change.

The Fork in the Road

One day, Doug realized he'd had it with his supervisor. *One more demand like that and I'll show him. I'll quit*, Doug grumbled to himself.

Sure enough, his supervisor asked him to cancel his other plans and fly to Florida the next day; this was an emergency.

"I'm sick and tired of your demands! What do you take me for—a fool?" With that, Doug stormed out of his supervisor's office and out to the parking lot. He sped out of the driveway, and within two blocks a police officer pulled him over and began writing out a ticket for going 50 in a 35-MPH zone.

As Doug sat in his car, it dawned on him that he was totally out of control and definitely in an unhealthy way of thinking. Once again, he had made an impulsive, reactive decision while in a rage. All of his musings about leaving the job seemed to build up to this point. *What have I done?* he asked himself. *Is this really what I want, or am I overreacting?* Suddenly, it became absolutely clear to him that he had been "temporarily insane" and had acted on that insane thinking.

When the officer handed him the ticket, Doug thanked him for the wake-up call. Puzzled, the officer walked away wondering why anyone would be thanking him for a ticket.

Doug suddenly realized that this was just one of numerous occasions in which he had set himself up for having a major emotional reaction by thinking that "they" were ruining his life. It became clear to him that he was taking everything personally and that it was his thinking that was giving him his emotional reaction. A flood of memories and insights washed over him, and he saw this pattern throughout his whole life, with his family, his bosses, his girlfriends, other jobs. It was always "their fault," and his only recourse was to get away from those negative influences. That had appeared to be his only option this time once again—up to the moment he got the speeding ticket.

Doug went back to the office and apologized to his boss. "I'm really sorry I reacted to you the way I did. I was way out of line. I'll be in Florida tomorrow." With that accomplished, Doug actually began to feel good about the trip and even decided to throw in his golf clubs and take the weekend off for some R & R. *Man, I'm so lucky to have a job that will get me out of the cold winter and off to Florida,* he thought, as he went whistling out the door.

On the plane back from Florida, Doug felt a warm glow of grateful feelings as he realized how much he actually loved his job, especially now that he realized where the power over his anger and his happiness truly resided. *Boy, am I glad I didn't quit. That was a close one. I wonder in how many other areas of my life I've been doing the same thing,* Doug mused. *I guess the grass only looked greener because I needed a new pair of glasses!*

The Principle in Practice

With the wake-up call of the speeding ticket and the insight that his experience was created in his own thoughts, Doug regained his freedom and his responsibility for his own life. He realized that the source of his happiness was inside him all along and the buttons that "they" were pushing to "make him angry" were being pushed by Doug himself from the inside out. When we realize the power of thought to create our emotions, we don't have to give up what we love to do in order to find our happiness.

Doug's story brings to mind the last line in the *Wizard of Oz*, in which Dorothy tells Glinda the Good Witch what she has learned from her adventures. "Well, I think that it wasn't enough just to want to see Uncle Henry and Auntie Em. And it's that if I ever go looking for my heart's desire again, I won't look any farther than my own backyard, because if it isn't there I never really lost it to begin with. Is that right?"

Glinda says, "That's all it is."

Happiness is inside.

2

PLAYING IN THE ZONE

Kay was shocked and somewhat humiliated when she was defeated by a woman in her tennis club named Sally, who was far less skilled at the game than Kay.

After the game, one of Kay's teammates commented, "Wow, Sally must be a rising star in the tennis club to beat you."

"No, she's really not that good," bemoaned Kay. "I beat myself. I was really in my head instead of on the court."

Kay is a very competitive athlete. She is an excellent swimmer and windsurfer and, for her age group, one of the best tennis players in her area. Last year she was at the peak of her game, winning several championships. The ironic thing was that she didn't really try hard. She just had fun.

Then, on her last birthday, she read her tarot cards, a sort of annual ritual. The cards said she was going to have an exceptionally good year. *That means I could do really well in the tennis competition this year,* she thought. *I'll double my efforts, practice harder, and really apply myself. After all, last year when I was just goofing around, look how well I did.*

And try she did. She got out all her tennis videos and books and studied up on her techniques. She practiced hard and really took it seriously. She quit having that glass of wine the night before a match, ate healthier food, and made sure she was in tip-top shape.

Rather than relaxing the night before the big annual tournament, as she would have done in the past, Kay hit the tennis books and pored over them like she was cramming for a final exam. The next day she confidently thought, *I'm really gonna kick some butt today!*

During her match, she was continually trying to remember what she had read and was analyzing every stroke. For some reason her timing was off, and her head seemed to be getting in the way of her performance. She was very tense and really didn't enjoy herself at all. And her game reflected this.

The Fork in the Road

After losing to her less-accomplished opponent, Kay was struck by a thought. *When I wasn't trying so hard, I did better than when I doubled my efforts. What's the deal?*

Then the answer hit her. *The tarot cards did say it was going to be a great year, and they were right. I just learned one of the most important lessons of my life. When I'm enjoying myself and not so serious, when I'm not analyzing every move I make, my performance is far better. That's what they mean by being in the zone!*

The Principle in Practice

Kay learned an important lesson. When her thinking on the tennis court was laborious and analytical, she made mistakes, lacked creativity, and was unresponsive to her opponent. And this applied to more than just her tennis game; she realized this across the board in all areas of her life. Whenever she relied on her intellect instead of her intuitive, "in the zone" thought process, life was less fun, she was less effective, and she wasn't as happy. When she was living in the zone, all her techniques and skills came to her effortlessly, whenever she needed them.

We are always thinking, but the quality of our thinking is constantly changing. When our thinking is serious, analytical, and labored, it becomes narrow, habitual, and less spontaneous. And when we are being analytical, it feels like we are *intentionally* thinking. But when our minds are relaxed, we are doing nothing more than *allowing* our thoughts to flow. In this relaxed state of mind, the right thoughts seem to spring forth from out of the blue, just when we need them.

When you respect this flowing thought process that has come to be known as "the zone," you get out of your own way and experience the zone's natural responsiveness to all life situations. Understanding the power of your mind to create this type of thinking is the secret to being a successful athlete, or even just having a happy life.

Kay recalled an interview with Andre Agassi, the former U.S. Open champion known for his talent in returning difficult serves. When asked how he could be so consistent, Agassi replied that when he was in the zone, the ball looked like a watermelon.

Playing and living in the zone are natural and easy once you learn how your mind works and how thinking creates your experiences. All you have to do is trust it.

3

SURVIVING A MERGER AND DOWNSIZING WITH EASE

Ginny has worked as a lab technician at the same hospital for twenty years. When her hospital merged with another hospital five years ago, the atmosphere at work became tense and uncertain, hostile and territorial, and filled with fear. To help employees cope with all the changes and facilitate the integration of the two hospital cultures, management hired a consulting firm and offered seminars to employees on stress prevention and how to deal with change effectively.

Three years ago, the changes continued as each department was asked to implement more budget-saving measures. In her small department, the manager, assistant manager, and her supervisor all lost their jobs on the same day. Ginny and her co-workers felt angry, abandoned, and full of uncertainty. Who would make the decisions? Who would determine the workload? Who would they go to when they had a problem? These were the questions that plagued Ginny and her co-workers.

About that time, one of her co-workers, Jeff, attended one of the hospital-sponsored seminars on stress and change. He came back from the seminar very enthused and excited, as he always did from seminars or vacations. As usual, Ginny and her co-workers ignored him and waited till Jeff got back to normal. Jeff initially hung up signs in the work area with sentiments like "Thought creates reality" and tried to tell his co-workers how easy it would be for them to change their attitude. They weren't interested.

Over time, Jeff realized that he needed to just put into practice what he had learned and not concern himself with his co-workers, no matter how valuable he might think what he'd learned would be for them too. He realized that his mental well-being didn't depend on their changing along with him. As a result, Jeff started to mind his own business and quietly changed his own attitude. He would catch himself about to have an emotional reaction to the latest pronouncement from management or to a potential conflict with his co-workers and would keep his mouth shut, leave the room till he got his bearings, and come back calmed down.

Ginny and the others wondered if he left to take a "happy pill," because his mood would change so abruptly. When he did talk, he usually had something very wise and comforting to say that helped calm them down too. As Ginny told me, "He was helping all of us to become calmer, more confident, and more of a team, and we didn't even know he was doing it."

After a while, Ginny saw that Jeff's new attitude wasn't wearing off. She had worked with him for almost twenty years and thought she knew what he was like, but Jeff became a calm and happy person instead of the aggressive, negative, and cynical guy she knew. *If Jeff can change this much,* Ginny thought, *maybe I should learn what he learned.*

The Fork in the Road

Recently, Ginny attended one of the hospital-sponsored seminars. All of the ideas Jeff had been sharing with her and exemplifying started to fall into place as she listened. Ginny realized that her strong emotional nature came from not recognizing the source of her own experience—her own thinking. She now sees why it is important to calm down first and then act, decide, or react. Her habit was to fly off the handle at the least provocation and then have to clean up the mess later.

After the first night of the seminar she shared what she was learning with her two teenagers. They said, "Mom, we've been trying to tell you this all our lives. You need to just chill out!"

She hated to admit it, but they were right. Humbly she said, "I know what you're saying is true, but before I didn't understand that I had a choice. It looked to me like you kids just didn't understand, that when you had the responsibilities that I do and worked at a real job, you'd understand why I react the way I do."

Ginny's department has changed dramatically for the better. Even though they have two fewer technicians and three fewer managers, they are more efficient now than ever. They are able to get as much, if not more, work done with fewer people. They used to have to live with decisions their boss made that were often irrelevant and not based on what their jobs really entailed. Now, they were making all the decisions themselves as a team and really knew what variables they were dealing with. Somehow they worked out the schedule, workload, and other details without much effort. It all seemed like common sense. The atmosphere in their department is now calm, cooperative, and there is a feeling of truly providing a valuable service to patients and staff at the hospital.

Ginny shared with me how she feels about all of these changes. "I worried it would go away, but it just keeps getting deeper for all of us. At the seminar I wanted to stand up and share how this has impacted me, but I knew I would start crying and not be able to say anything. Not because I am sad, but because I am so grateful. I can only compare the feelings I am having to when I gave birth—total joy."

The Principle in Practice

I recently spoke with the director of the Learning Department at Ginny's hospital. She felt that Ginny's story exemplified the kind of change that management was hoping for when they committed to improving the human-relations climate at the hospital. Convincing the management of the hospital to commit to this type of long-term program was difficult initially, but the results have paid off in terms of a rise in morale, communication, and productivity. The hospital has now expanded its program to the whole community and is sponsoring seminars for government, businesses, and schools.

Applying the principles of mental well-being to organizational change requires a commitment to a more long-term attitudinal change and not just a quick fix. Organizational change occurs one person at a time, as it did with Jeff and then Ginny. However, when one person changes, it gives hope to all those around him or her. Even before Ginny changed, Jeff's calm attitude and wisdom had a positive effect on their work group. Their meetings were more lighthearted and productive, and it has become easier for them to come to an agreement on decisions. Once a critical mass of attitudinal change occurs, there is a shift in the entire organizational culture.

Human beings have an innate desire to feel happy. Many have given up hope that it is possible to be happy, especially at their jobs. Realizing that happiness comes from an internal source rather than from our boss or co-workers is a revelation that can free us, like Ginny, from blaming the tyranny of the organization for our state of mind. Things like policy decisions, mergers, and downsizing may be out of our control, but we can control the most important thing—our own mental well-being. We can do this, as Jeff and Ginny did, when we gain an understanding of our mind.

Positive and permanent organizational changes occur one person at a time.

4

BRINGING ENJOYMENT TO WHATEVER YOU DO

So many of us are in the habit of delayed gratification—we wait till we're done with our list of to-do's before we allow ourselves to take pleasure in life. When we believe our happiness, self-esteem, and worthiness to enjoy life are dependent on accomplishment, our work and tasks become drudgery. We can choose, instead, to enjoy everything we do, no matter how insignificant.

I met Liz at one of my book signings. An elderly woman, Liz has been on a quest for peace of mind for much of her life. She had read innumerable self-help books and studied meditation and other techniques. During my presentation, I mentioned that anything in life could be enjoyable, even doing the dishes, because all events in life are neutral—it is only our thinking that judges them as good or bad.

After I made that comment, Liz excitedly raised her hand, "I was with you up to this point, but you've gone too far with that last statement. I can't imagine myself ever enjoying doing the dishes. I hate doing dishes! I loathe doing dishes! How can doing something as mundane as that ever be pleasant? If I can ever learn to enjoy doing dishes, I'll know this new method of yours really works."

Immediately, someone else in the audience stood up and said to Liz, "I love doing the dishes. It's almost like meditation for me. I let my mind wander and dream, and I even take my time doing it because it is such a pleasure to me."

Liz retorted, "You must have lost touch with reality! No one in their right mind could possibly like doing the dishes." She shook her head in disbelief and disgust.

At the time, I thought Liz was quite a character, but I wondered, given her level of conviction about the inherent drudgery of doing dishes, whether or not she would ever change her mind. Then I started getting phone calls from Liz. She wanted to learn more, she said, and asked if she could come to my next seminar. I told her she could, but I didn't imagine that she would actually attend.

Sure enough, on the first day of that four-day seminar, Liz was there in the front row, eager to learn more at age eighty-one. I admit I was a bit intimidated when I saw her, because of the number of questions she had asked at my book signing. I wondered if she would disrupt the group. To my surprise, Liz became such a hit that at the end of the seminar, I jokingly told her I would pay her to attend all of my seminars because she contributed so much with her humor and her wisdom.

"I want to give an unsolicited testimonial," she said with what I came to know as her characteristic dry wit.

Fearfully, I nodded for her to proceed.

"All my life I've hated doing dishes. When Joe said everything in life was neutral, only our thinking makes it good or bad, I was sure he was wrong about one thing—dishes. I could see it made sense for the bigger things in life, but in my mind washing dishes was like cleaning the toilet, inherently unpleasant. But I decided to keep an open mind and have been applying the principles of slowing down to everything in my life. One day, I was dreading doing the dishes and decided to give it a try. I decided that every time I had a negative thought about doing the dishes, I would just see it as my lifelong habit and consider that maybe I was wrong. I opened myself up to the possibility that even at age eighty-one, with decades of experience hating doing the dishes, I could be wrong. After all, even my husband was looking sexier these days! If he could go from looking like a shriveled-up old man to a sexy senior citizen, maybe the dishes could change too!"

The Fork in the Road

"There I was, standing at the sink," said Liz, "when suddenly I realized how much time I had wasted all my life being miserable till I got the dishes done. And my brief enjoyment at having completed that task would always be short-lived, because my husband immediately started dirtying more dishes and the dread would begin all over again. I started seeing that the little act of doing the dishes had robbed me of perhaps a quarter of my life of happiness. How silly of me to have been so unknowingly tricked by my habit of dreading the dishes! I burst into laughter and have enjoyed doing the dishes ever since."

The Principle in Practice

Just as Liz used to feel about doing the dishes, most of us have tasks that we dread and can't wait to complete so we can get on to the fun things. And so we end up delaying our enjoyment until after we finish the hated task. At some point in time, we have all innocently learned or assumed that some things are inherently unpleasant and are therefore stressful, awful, or disgusting. We've come to accept that there are just some things in life that have to get done whether we like doing them or not, and the only joy is in getting them over and done with. The result is that without realizing it, we rob ourselves of many hours of enjoyment. We live in misery—all from a thought that *we* made up.

What Liz realized was this: *It was her habitual way of thinking about doing the dishes that was unpleasant—not the task itself.* Liz had thought that her dread came from the inherently distasteful task of washing dirty dishes. It had never dawned on her that her dread had anything at all to do with her thinking.

Three things created a shift in Liz's perception: (1) she wanted to be happy; (2) she was willing to consider that her happiness (or unhappiness) was the creation of her thinking; and (3) she witnessed her thinking. When we, like Liz, are able to recognize our thinking in the moment—even if that thinking is a lifelong habit—it loses its power to take away our happiness.

What tasks in your life still rob you of happy moments until you get them done?

Take the drudgery out of life—clean up your thinking!

5

POSITIVE THINKING ISN'T ENOUGH

Jerry is a very successful motivational speaker and business consultant. Now in his forties, he has spent the last twenty-odd years perfecting his skills and learning everything he could on self-help and success, including going to every positive thinking and motivational rally—from Anthony Robbins's to Stephen Covey's. For the past ten years, he has also been an ardent student of the principles in this book. He has always considered himself "Mr. Positive Thinking" and is considered successful by most standards.

Last fall, Jerry began experiencing a serious health crisis that would ultimately change his life. He had suffered from ulcerative colitis for many years, but this bout really brought him down. Bedridden for several months, he became totally dependent on his wife. He couldn't even get back in bed or onto the couch without help.

Frightened about his condition, Jerry asked his family physician what he could do. His doctor said, "Jerry, unfortunately I don't see a very positive prognosis for you. People with your type-A personality don't really change, and I'm afraid as long as you stay the way you are, you will be prone to this disease. Your colon is reacting to the stress of your lifestyle, and you'll probably lose your colon someday or die from colon cancer."

Horrified and shocked by this dismal pronouncement, Jerry thought, *Oh my God! What have I done to myself? How could I have let this happen? I know how to be happy and calm. I teach it to others better than almost anyone else. But I know it's true—I've made myself sick with my own frenetic lifestyle and speeded-up pace.*

As Jerry reflected further, he realized that for years his doctor had been telling him that his stress was the cause of his colitis and his bleeding gums, but he had always refused to believe it. In fact, he was always offended that his doctor would even make such a suggestion to "Mr. Positive." Suddenly, he realized that he must have been in big-time denial.

The Fork in the Road

As he lay in bed for months, however, Jerry's mind began to slow down and a new sense of calm began to set in. He shared with me what happened.

"After all that time in bed, my thinking caved in on me," he said. "I finally started to slow down. I liked the new pace of my life, though I didn't like all the pain that was forcing me to slow down. But I began to see that this illness was my wake-up call—a blessing in disguise. Once I realized this, I vowed to make a calmer pace a priority in my life. A calm life would become every bit as important as success. As I lay there I began to think that I should do a 'Jerry Seinfeld' and bail out while I was at the peak of my career, rather than on the downturn.

"Once I was healthy enough to go back to work, I began turning down business and limiting my speeches to one and a half hours, but keeping my fee the same. To my surprise, no one complained and I actually had more business than before with a smaller percentage of the effort. Suddenly, I realized I could have success and a calmer life too. It dawned on me that my own beliefs had limited my success and forced me into a crazy lifestyle that was destroying my health. Now my only regret is that I didn't realize this ten years earlier."

In retrospect, Jerry realized that his stress had been invisible to him. He had disguised his stress as something positive—he had called it passion, focus, intensity, all of the buzzwords of the motivational world. And he had fooled himself into believing that living with stress was necessary if he wanted to succeed.

"I had understood the concepts of mental well-being before," he explained to me, "but only on an intellectual level. It was the shock of being ill that cleared my head, and suddenly I got X-ray vision for the principles. I really started to truly understand them.

"Recently, I got a call from a customer who said, 'Jerry, I've got a deal for you that could make you a lot of money.'

"I began to salivate like one of Pavlov's dogs, but then I realized that old familiar feeling. I started to get revved up, and my whole body got an adrenaline rush. Then I remembered what I learned about listening to my feelings and sensations as a signal, that this is my body trying to tell me that my thinking is getting speeded up. Sometimes I can see this is happening right away, and other times I can't. But if I don't see it immediately in my thinking, it will show up in my gut and my digestive system. That's when I realize that this is my signal to back off. Then I listen to myself and reflect on what to do rather than react out of habit.

"I also feel now like I'm getting my conscience back. My mind was so busy that I didn't notice what other people around me were feeling—or how I was impacting them.

Jerry notices now that in certain situations in which he used to become angry, upset, or judgmental, he feels compassion instead. Recently, he was on an airplane and told me about an experience he had.

"Normally I fly first class," he said, "but it was filled, so I had to sit in the back of the plane next to a man with his young daughter sitting on his lap. She cried all the way from the beginning to the end of the flight. Usually I would have demanded that the flight attendant trade seats with me, but this time my heart went out to the father. In the past I had an intellectual understanding of compassion, but I didn't actually experience it. But this time, the situation didn't bother me at all. The man was very apologetic, and all I could do was imagine how bad it was for him. This was really new for me."

The Principle in Practice

Since Jerry has broken out of the speed trap, he has found not only his conscience, but the deeper and more positive feelings that come with a quieter mind. He used to merely preach about mental health and what it was like to live in a state of well-being, but now he truly lives it and is reaping the benefits of his understanding.

There is much that we can learn from Jerry's story in order to experience our own health realization. Through his illness, Jerry's body was trying to point him back to his innate mental health, as it does for all of us. When our mind gets out of balance with worry, stress, anger, or any other negative emotion, it feels uncomfortable to us. If we ignore that discomfort for long periods of time, it becomes chronic—and often turns into disease. If we treat the disease's symptoms but ignore the message it is giving us, it will persist, as did Jerry's ulcerative colitis. So much of disease is innocently self-created and self-perpetuated, simply because we fail to recognize that uncomfortable feelings and sensations are all signals about something being out of balance—our diet, our lifestyle, or *our thinking*.

Jerry was fortunate enough to recognize the basic principle that thought creates experience when his own wake-up call arrived.

Disease is the body's way of communicating to us that we are out of balance in some way. When we understand the principles of mental well-being, along with good nutrition and exercise, we are able to truly be an active participant in the prevention of illness. Unfortunately, most people are totally unaware of how to change their level of stress other than to cope with it or temporarily alleviate it. Armed with an understanding of the principles of mental well-being, you will be empowered to regain your full health, as did Jerry.

Listen to your body. It is connected to your thoughts.

6

BREAKING THE WORRY CYCLE

It seemed as though everything in Mark's life was going through change and dissolution. He was in the midst of a highly contested divorce, and, partially due to his depression and inability to function, his business partners decided to break off on their own.

Mark's whole approach to life was to put everything into a category or box. As a financial/estate planner, Mark excelled at analyzing balance sheets and investment portfolios and making projections based on his analysis. Now he found himself analyzing *his* future—facing life without his wife, the question of how he would relate to his children, and financial uncertainty from the breakup of his partnership. As he obsessed on these issues, his mind would spiral out of control with worry, fear, and depression. Some mornings he had a difficult time getting out of bed and getting dressed. He felt immobilized.

Mark was in a state of panic much of the time from weighing the outcomes of all the possible scenarios. He saw a traditional therapist who helped him analyze his past, how he got where he was, and how to cope. He was on a high dose of antidepressants and was even hospitalized for depression. Still, he wasn't getting any better; in fact, he was getting worse. The more he thought about the mess he was in, the more he spun his wheels, and the more he felt depressed. For Mark, there seemed to be no way out.

The Fork in the Road

At the urging of his former business partner, Mark changed therapists to a counselor who focused on teaching healthy principles of mental well-being. She saw Mark going down into a vortex of life changes without any understanding of how his thinking was playing into it. At first, her approach sounded too simple for Mark, and he had a hard time letting go of his habits of analysis and worry.

"I was on my way to another one of my divorce hearings," said Mark, "and I was listening to a tape about breaking the habit of worry. Suddenly, all of the ideas my therapist had been trying to get across to me fell into place. It was like a light bulb went on. Something changed, and all of a sudden my mind cleared up. I began to see that my habit of worry was nothing more than that—a habit. I saw that I worried my way through life, so no wonder I was stressed and depressed!

"My whole life began to change. I am still going through a difficult divorce, but I don't react nearly as much as I did before. I don't spiral down into the 'poor-me's' and waste time trying to figure out why this is happening. I just deal with one thing at a time, and that is manageable. In my work, I still need to analyze at times, like analyzing someone's estate tax liability. That's okay, but I'm able to recognize when intellectualizing is no longer being productive. I am now able to put things on the back burner and let them go. It is amazing how creative my thinking has become and how much the quality of my work has changed. When I let go, the answer just comes. And, of course, I am far more productive now. I'm also able to sleep."

"How about your feelings of depression?" I asked.

"I still have times when I feel depressed or overwhelmed. But now I only feel that way for an hour or two instead of weeks at a time. I'm able to recognize my thinking and recognize that I'm in a low mood. I used to spend so much time trying to control outcomes that were all in my imagination. Now I save so much energy not doing that. As soon as I recognize that I'm trying to control things, I stop it right away. I'm able to question my own thinking instead of trusting thinking that is isn't based on reality. I haven't taken antidepressants for several months now, and I feel better than I ever have.

"I believe that if I hadn't learned these principles, my life would be totally different. I've been able to turn my business around, I've lost over sixty pounds by following an exercise program, and I'm completely motivated to live my life. I don't know if I would have even been alive if I hadn't learned this. My biggest realization is that life doesn't fit into a box anymore. There are endless possibilities now—I'm volunteering to be on boards of nonprofits, I listen to people now, I get up each morning to watch the sun rise, and I am awestruck with every day and what it has to offer. I can't believe that life can be so much fun and so joyful."

The Principle in Practice

Worry is a habit for many people. I believe that worry is an attempt to deal with the fact that much of life is an unknown. If you don't have faith that life will work out for the better or that you will be able to deal with what comes up, you have no choice but to worry—to process and analyze all the possible scenarios.

When you begin to trust in the power of wisdom and insight, when you have an understanding of a larger force at work in life, you can let go and recognize when you have gone overboard with your future projections.

Mark was used to controlling life and putting it into explainable categories. When his divorce and his business problems got to be too much for him, his old coping mechanisms no longer worked. Luckily for Mark, he was able to learn a deeper, more profound way to deal with the unknown in life. He was able to recognize when his thinking was spinning out of control, before it led to exhaustion and depression. Now he is able to put things in perspective and see the boundless possibilities that life has to offer.

By trusting in his inner intelligence, his wisdom, Mark has found that he can handle even the most difficult situations. He isn't perfect; he can still get caught up in his thinking and his low moods, but his lows are short-lived. He now sees analytical thinking as a tool for certain tasks, not a weapon to be used to judge himself and others or a means of figuring out the unknown.

The future is unknown. Instead of worrying about it, have faith that your wisdom will guide you into the future.

7

"MOM, THERE'S A STORM MOVING THROUGH ME": ONE PARENT'S APPROACH TO ADOLESCENT HORMONES

Judy wondered what the adolescent years would be like for her son, Dan. When he was five years old, she and Dan's father were divorced, and since then she had raised Dan as a single parent. Nonetheless, Judy had found Dan to be an absolute delight to be with and to parent. Her family, neighbors, and friends would often warn her, "Just you wait till he's twelve or thirteen . . ." He was now fourteen, and Judy still saw being his parent as the greatest pleasure of her life.

Because her own teen years were a nightmare, Judy vowed not to repeat the pain she and her parents went through. Fortunately for her, just before her son's first adolescent hormone attack, she had learned about how her mind worked.

It was a night just like any other. Judy asked Dan to pick up his socks. Dan ran to his room and, in a moment, Judy could hear sobs coming from the bedroom. Cautiously, she walked into his room and sat on the edge of the bed. Dan said, "I don't know what's happening to me, Mom. I don't know if I'm angry or sad or what's going on. What am I feeling? What's happening to me?"

The Fork in the Road

Normally, Judy would have tried to reason with him and find out what was bothering him or tried to talk him out of it. But she knew from what she had learned about her mind and her thinking that she should just listen and be quiet. She felt compassion for her son, but knew better than to reach out and hold him.

"I needed to just let it pass," she later shared with me.

After a while Dan said, "Whew! That was awful, Mom. It was like a storm passed through me."

Now that Dan was calmer, Judy said, "Now, here's the plan, Dan. You're approaching adolescence, and I don't know how often this will happen to you or how long it will last, but it will happen from time to time. This is how I propose we handle it. Hormones are going to go through your body, your moods will change suddenly, and you'll have lots of different emotions. There's nothing to be afraid of or concerned about. It'll be happening on and off for years. When it happens, I'm not going to try to talk to you or reason with you. I'll just wait it out, because I know that it's not really you—it's just your hormones going through your blood. After it's passed, if there's something to discuss we'll discuss it, or if not we'll just move on."

Dan was clearly relieved that his mom didn't react to his outburst with one of her own or take it personally and that she seemed to understand. It also helped him that she explained what was going on.

"Okay," he said, kind of trembling. Then Judy hugged him and told him she loved him. They would get through this, she assured him.

This common incident could have turned out quite differently, as it does with many parents and teenagers. Many parents try to argue with their kids or reason with them. Or they take it very personally and feel hurt. Or they may feel that their authority is threatened and try to gain control. All of these options end in bitter battles that can last for many years and often permanently damage their relationships.

During the period in which she was taking one of the Health Realization courses, Judy had a second experience with Dan. She arrived home tired after a long day at class, but very exhilarated with the experience. She expected she would have a nice quiet evening with Dan. Then round two occurred.

"Mom, I have to go to the barber *now*," Dan announced with great intensity.

It happened to be just before the barber shop closed, and Judy didn't think it was a matter of life or death for Dan to have his hair cut that evening. So she said firmly, "No, Dan, I'm not rushing you over there to have a haircut. You can wait till tomorrow and we'll make an appointment."

"Why not? Don't be so lazy, Mom," Dan was screaming now. "All you ever do is sit around! I want it cut tonight. Look how long it is! I look like Don King!"

Since Dan's hair was about a half-inch long, Judy could see that the situation was really getting absurd. But she kept her comic perception of the scene to herself and continued to listen to him quietly. However, she began to doubt her strategy when Dan's extreme mood persisted for several hours. This had never happened before. Still, she stuck to her original plan and didn't try to reason with him. He was being rude and angry and quite unlike himself. In the past, she would have been tempted to yell back and send him to his room or maybe punish him in some other way. But again, she just got very quiet and listened to him, hoping the mood would pass soon.

After a couple of hours, Dan took a deep breath and said, "Boy, I'm glad that's over. That's fine, Mom, I can get a haircut tomorrow."

The Principle in Practice

Unlike most parents faced with an adolescent temper tantrum, Judy handled both of these situations in a way that prevented them from escalating. What Judy knew and did, which was to stay calm and just listen until the mood inevitably passed, kept her from falling into the common parental traps of arguing back, overreacting, taking it personally, becoming judgmental, getting angry, and saying hurtful things.

From Judy's story, parents can learn five key approaches to dealing with adolescent mood swings:

1. *Recognize moods and know that they will pass.* Judy can recognize when Dan is in a mood and knows it would be foolish to

focus on the content of what he is saying or doing. Instead, she waits out the storm and knows it will eventually pass.

2. *Don't take it personally.* Judy knows that even if Dan is being disrespectful or argumentative, it has nothing to do with who she is or her capability as a parent. Her self-esteem isn't on the line.

3. *Keep a sense of humor.* If there is one quality necessary for raising an adolescent, it is having a sense of humor. It's either that or go crazy.

4. *Trust in your child's innate common sense.* All too often, parents remember the foolish things they did in adolescence, or they question their child's judgment based on his or her moods. This causes parents to doubt the common sense of their kids. Most of the time, our children are on their own, and all they have to guide them is their wisdom. If they are to mature, they must learn to trust their wisdom and listen to it. Reinforcing your belief in them and their common sense is one of the greatest gifts you can give your children to help them prepare for life.

5. *Listen.* The number-one complaint I hear from kids is that their parents don't listen to them. Listening allows your child to feel respected, it gives you time to calm down, and it allows time for their own insights to bubble to the surface. Listening is the WD-40 of relationships.

Parenting is one of the most daunting tasks we face in life. Following these simple guidelines could save you and your child a lot of heartache and provide a solid basis for your relationship with your child in the years to come.

Remember, in dealing with your children, listen, know this mood too shall pass, trust in their common sense, don't take it personally, and keep a sense of humor.

8

A PARENT'S DEATH: THIS TIME WITHOUT FEAR

When Sue's father died about twenty years ago, it was a traumatic experience for both her and her father. He was a very athletic man who was healthy until he found out he had cancer. Over the next six months, Sue watched him deteriorate to the level of an infant. She had to parent him and take care of all his needs. All she felt was fear; all she wanted to do was run.

Twenty years later, Sue had matured a great deal. She had been on a search her whole life for happiness—trying to overcome years of depression, anxiety, and alcoholism. She spent a small fortune on therapies, most of which seemed to take her deeper into the past. Still, her problems worsened. Then one day her therapist announced that she had learned a revolutionary approach to counseling that would help Sue. As Sue watched her therapist transform before her eyes because of this new approach, Sue began to feel hope for herself. Her therapist was now more light-hearted, hopeful, and full of life. As Sue learned the principles of how her mind worked in a healthy way from her therapist and from workshops she attended, she too transformed her life. She no longer felt the heavy burden of the mental illnesses she had been labeled with and, instead, began to identify with her core of mental well-being.

Finally, she thought, she could enjoy her life. It was going to be great! She went on a vacation to Vancouver with her mom, and they had the time of their lives. One night shortly after they returned, her mom called and complained of a pain in her glands. Because it was unlike her mom to complain, Sue insisted she see a doctor. After many doctor visits and a biopsy, Sue's mom was diagnosed with cancer.

Sue was terrified. The memories of her dad's death flooded her mind, and she fell into a panic. It wasn't fair, she lamented. She had just gotten her life together, and now she would have to go through this all over again. She became absorbed in self-pity.

After reflecting on the news, Sue returned to her mom's hometown to be with her. She became determined not to repeat the terrible experience she had had with her dad's death. Throughout the next two years of operations, chemo- and radiation therapy, recovery, relapse, and death, Sue was pleased to discover that her knowledge of how her thinking and her moods functioned kept her deeply grounded and calm. She knew there was a way to go through this experience with calm, regardless of the outcome.

Sue decided to make the time she had with her mom positive in every way possible. She took every opportunity to do things with her that were special to them both—take walks together, go shopping, watch the "Golden Girls" on TV every night. Most importantly, she didn't run away emotionally from her mom because of her fear. For the whole two years, she was *present*. She lived each moment fully. When her mom got angry or lashed out at her when she was in pain, Sue didn't take it personally. She felt compassion and patience instead.

Little by little, Sue taught her mom what she had learned about her moods, her thinking, and her wisdom. The compassion she felt for her mom began to be reciprocated, and it grew between them throughout the two years. The love they felt for each other was overwhelming.

The Fork in the Road

Ten months before her mom's death, the doctor gave Sue the news that the cancer had spread to her mom's bones, and she had six months to a year to live. Sue knew she had to tell her mom. One night they were lying in bed watching the "Golden Girls" when her mom said, "Sue, I don't want to go on anymore. I don't want to suffer anymore."

"Don't worry, Mom. You don't have to," Sue said softly, and then explained the doctor's prognosis.

Sue began to cry, but noticed that her mom was elated by the news. She looked radiant. She was very relieved to know there was an end in sight; she was ready to die. Her mom's elation woke Sue up too, and she decided to feel good about it as well. During the next ten months, they healed a lifetime of pain in their lives by sharing love, joy, and laughter. What helped Sue most was seeing the total innocence of her mom's approach to how she had lived her life. Her mom had lived her life as she perceived it through her thinking—life was a difficult struggle to be endured. All she could feel was compassion and forgiveness for herself and for others. For the first time in their relationship, they truly *listened* to one another deeply, without judgment or trying to change each other. Instead of being self-absorbed in her own sadness or fear, Sue was available to go through the experience fully present and awake to every moment.

The Principle in Practice

When Sue shared her story, she said, "The principles of mental health were the glue that kept me together, no matter what life presented to me. Going through my mom's death and knowing that my core of health was within me allowed me to quiet down my thinking and to live without fear. I realized I had a choice—I didn't have to go through the death experience with my mom the way I had with my dad. I could either be full of fear and anguish or in a state of calm and love."

When we have a foundation of mental health and know that it is always within us, we can be resilient regardless of circumstances. Going through the death of a parent is certainly one of the most challenging issues many of us face as adults. There is no way to avoid the grief and sadness of a loss as great as this. However, if we go through it while staying in touch with our core of health, as Sue did, we can cherish the moments we do have together. We can heal our relationships if we need to and feel compassion for the dying person and those around him or her.

Dealing with the death of a loved one is one of the most potentially growth-enhancing and loving experiences we can go through. It can also be a time of division and fear. When we trust in the power of our mental health, our wisdom will guide us through this experience and fill us with a sense of gratitude and love. It will also help prepare us to face our own mortality.

When you discover the core of mental health that is within you, you can even go through the death of a loved one with feelings of calm, love, and compassion.

9

A BUSINESS FINDS SUCCESS THROUGH HAPPINESS: A CEO'S STORY

Six years ago, a large national advertising company found itself in a serious financial crisis. Although it was a relatively successful company, it was overly financed and was failing to service a $125 million loan. Its lenders were giving the company six months to turn the business around or they would assume control of their assets. Several of the owner's other businesses had already been sold to pay off debt, and this was the next one on the chopping block.

The mood in the company was anxious and near panic. The employees were trying hard to maintain their equilibrium in spite of the financial burdens, but it wasn't easy. They were putting in long days and nights, but seemed to be losing ground despite their efforts.

One day the CEO got a call from the company's owner. "Kurt, I'd like you and your regional managers to go to a wellness workshop next weekend, and bring your spouses. Please have our CFO make the necessary arrangements. I think this can help pull us out of this slump."

That was all he said, and Kurt was left feeling confused about how something like this could possibly help them with their current crisis. Nonetheless, he listened to the owner and put things in motion for everyone to travel to a small town in the state of Washington in search of "well-being."

Kurt and his managers arrived at the seminar feeling guarded and suspicious. They wondered if this was going to be some sort of confrontational group or the owner's diplomatic way of announcing the end of the business. Their imaginations were definitely out of control as they pondered the possible outcomes of the weekend.

Despite their worries, just being away from the pressure cooker of their day-to-day business situation seemed to help calm them down. They were in a vacation-like atmosphere with their spouses, and the seminar leaders didn't challenge their present business strategies. The consultant assigned to their group listened to them and basically felt that they were a very competent group of people. The seminar they attended focused on how to maintain a healthy state of mind in a difficult situation.

By the time the weekend was over, the managers and the CEO felt refreshed and more hopeful, yet they had no clue as to why they felt this way. Nothing had really changed—they had no new strategic plan in place for saving the company, but they felt calmer, had a greater sense of trust in themselves and each other, and felt more confident.

Before the group left Washington, their consultant asked to meet with Kurt again individually.

"That sounds like a great idea," Kurt replied. "Maybe two years from now. But right now I have a company to save that's on the brink of bankruptcy and I don't have time for myself."

"It looks to me like you have a choice," the consultant said. "You have to build a new foundation for this company if it's going to turn itself around. You can do that in one of two ways. You can do it from a healthy level, or you can do it in a way that will cause everyone to hate you. Your style right now is abusive and controlling, and your employees have a chip on their shoulder about you. In all fairness, however, I don't think you have a clue that you are the way you are."

Shocked, Kurt responded, "I'll see you next weekend."

The Fork in the Road

That next weekend Kurt discovered how his personality intimidated others under him in the company. He had always been told that his type-A personality was perfectly suited for a CEO position, but what he was learning now was the complete opposite. Kurt learned that his personal style was not only intimidating to his employees, but also to his wife and children.

The consultant shared an alternative approach with him that was based on creating a healthy work environment and building trust among the employees. Intuitively, it all made sense to Kurt, even though it appeared to go against common business practices.

Years later, Kurt shared his realizations with me. "I realized I had to make a choice between my habitual high-pressure, high-intensity way of doing things and a new way based on calm and wisdom. It was a leap of faith, but it felt right. I had an epiphany that weekend that was somewhere between the intellectual and the spiritual realms. I started to see that it was an inside-out world and that we were all creating our realities with our thinking. Responding with calm and wisdom seemed a lot better than reacting habitually to outside forces and pressures. Over the next few years, my employees and I learned how to operate more instinctually and in the moment with a creative approach rather than adhering to a strategic game plan.

"Our business is constantly changing and responding to the fluctuations in our markets. We have slowly evolved into a company that is happy as well as successful. Before, we were very unhappy—stressed, hostile, reactive. Now our employees genuinely enjoy their jobs. I don't think there are any employees who would say that their job is getting them down."

I asked him, "How does it help a company to create an atmosphere in which employees can be happy?"

"Several things happen," he said. "First, defensiveness comes down. People who are happy put other people at ease. Before, they felt fearful of making a mistake or saying something stupid. Now they trust each other and open up more, sharing their ideas in a creative exchange. As the atmosphere became less controlling, ideas began to flow and the focus was not on office politics and competitiveness, but rather on getting the job done. Motivation became an internal drive, rather than the result of promises of external rewards to employees. I've found that when people are internally motivated, they get twice as much done in half the time. My managers' sole job is to provide an atmosphere in which people can be happy. We don't feel responsible for making people happy, but we show them how to do it themselves. We don't have to baby-sit our employees anymore, which leaves us more time to actually do our jobs—to be creative in looking at the markets and seeing opportunities.

"Our meetings at management retreats used to last eight hours and were very detailed and tedious. Now they last two hours maximum, and they reach a profound level of discussion very quickly, getting to the heart of the matter. If you don't trust your employees, you have to go over each detail for fear they won't know what to do. But if you trust in the health of your employees, you see their common sense come out. The most important thing is that employees are no longer afraid to make mistakes. They know they won't be judged if they do. This allows them to take bigger risks with greater rewards."

"How does this translate into the bottom line?" I asked Kurt.

"We made a $567 million swing in our company's value in the past six years. We went from a deficit of $17 million to $550 million in assets. We more than doubled our 'same-store business,' that is, twice the income for an equal amount of overhead. One good idea is worth a lot of money. The other bottom line is that I work with a group of people who love their jobs, and I do too. That gives me enormous personal satisfaction. The financial rewards are the frosting on the cake."

The Principle in Practice

When Kurt had his epiphany while meeting with the consultant, his understanding of the source of his experience of life changed. He realized that the cause of his stress was internal and was not caused by the competitive nature of his business. Most important, Kurt saw that he had a choice: he could continue in the direction of his habitual type-A personality, or he could manage his company and his personal life from a new level of understanding that was based on wisdom and a calm mind.

When our level of understanding increases, we suddenly see more choices in our lives. Things that appeared fixed and unchangeable, such as what we might have considered to be a part of doing business in today's world, now seem like options instead. Other ways of dealing with life suddenly become obvious. To Kurt, it was clear that the choice of operating from calm was the new direction he wanted to take both for his own sanity and the atmosphere and success of the company.

Most business management operates on the basis of control, fear, external motivators, and a lack of trust in employees' abilities and goodwill. In contrast, what Kurt developed in his company was *an atmosphere and a business philosophy based on healthy thinking—calm, reflection, insight, creativity*. The faith in this type of thinking among management and employees results in internal motivation, good decision making, mutual trust and respect, and honest and open communication. When employees are not stressed, they think in a more clear-headed, wise manner. As the CEO put it, "A good idea is worth a lot of money." When employees are in this creative flow of ideas, they are more responsive to the changing forces in the marketplace and are able to come up with solutions to keep their business competitive with other companies.

Instead of having to hold employees' hands through innumerable personal problems and office politics, Kurt was able to put his energies into anticipating market changes before they even occurred. Most managers are in a crisis mode most of the time, so that by the time one problem is fixed, another one is already forming. Kurt now feels that he is ahead of the game and able to anticipate problems and solutions before they come up.

Kurt's company and its employees have realized success in two ways: financially and personally. He had the wisdom to see the value of putting his well-being and that of his employees first, and having faith that the financial rewards would follow.

Happiness lowers overhead and increases profits.

10

TRANSFORMING PANIC INTO PEACE

From the time Michelle was a young child, she was a caretaker. Her mom and dad were divorced when she was five. Her dad had emotional problems, and though she tried hard to make him happy, he was always depressed. Michelle's father wasn't her only problem; from age five through age ten, Michelle witnessed her mom's alcoholism. This was a prime situation for producing a person who would always be worrying, controlling, and trying to fix everyone else. Looking back, Michelle could see how these early years formed her adult personality.

Later in life, Michelle became a workaholic, trying to support both her alcoholic boyfriend and her parents. At one point, she almost went financially bankrupt trying to help everyone else out. She worked at a law firm, often twelve hours a day and seven days a week. She was so anxious and driven that she often suffered from panic attacks. At times she would become so immobilized with these attacks that she couldn't think and was very confused. Sometimes it would get so bad that she couldn't even choose what clothes to wear.

Michelle was an independent person who was used to figuring things out on her own, so the thought of seeking therapy went totally against her grain. She did end the relationship with her alcoholic boyfriend, but was still working ridiculous hours and completely pouring herself into her work. Finally, one day she called me, certain she was on the verge of a nervous breakdown.

"I think I need help," she said. "I can't sleep and I can't function in my job anymore. I'm having more and more panic attacks, and I'm afraid I'll lose my job if I don't get help. Do you think I need help?" She was almost apologetic.

I did my best to reassure her and said, "Michelle, I know this looks bad now, but this could be the turning point in your life if you get the proper guidance. I think you should see a professional counselor."

"I'll do anything at this point," she admitted.

The Fork in the Road

After that call, I didn't hear anything from Michelle for six months, but recently she shared her progress with me.

"At first I didn't understand what my therapist was trying to tell me," she said. "I felt angry at her because she knew something that I didn't know, and I always wanted to know and understand everything. However, when I was with her in her office, I would calm down, and I experienced a peace that I had never experienced in my life. I found myself not wanting to leave her office. Eventually, through the therapy and the tapes and books, I began to feel that peace on my own. My previous coping mechanisms of keeping busy and making to-do lists quit working and were replaced with my health."

"What turned you around?" I asked.

"My therapist told me to calm down. This was frustrating for me, because I had no clue *how* to calm down. Instead, I would ask her a list of questions each session and she would always give me the same simple answer—'Just let go and trust that you will know what to do.' Then she began to talk to me about how I had two ways to think, one was to process everything and keep it churning in my brain. The other was to let my thoughts happen and flow in and out of my mind. I found that when I did the latter, my mind became less muddled and things started to get clear. Before that, there was no room for new thoughts and no room for me to do anything with my thoughts. I had to learn to let go of my thoughts. I also learned that the content of what I worried about was meaningless; it was the actual habit of worry that was significant. This helped me understand why I had panic attacks.

"I began to be able to stand outside of my thoughts and see myself thinking. It became easy to catch myself thinking in dysfunctional ways. I learned to surrender when my thoughts got out of control. Strangely enough, I had to let go of my need to control in order to regain control. In a sense, my nervous breakdown was a gift. I finally discovered my health after all these years of coping. I found that when I just relax, my health is already there. There is nothing I need to do. It was like learning to float. At one point I had to just trust in myself and let go. When I did, I discovered that I could already float. I learned to just *be* and not *know*. Ironically, this is when my life started to feel back in control. Things just started coming.

"For example, I went back to school and took a programming class. I had always struggled with math and computers, but suddenly it was easy and fun. In the past, I had scared myself with thoughts of how I was going to do and the pressure of the final outcome. This time I just trusted in myself instead of my brain alone. It was a miracle. I don't have much stress anymore, and I actually get more done at work and come home earlier. I like life now, and I'm happy!"

The Principle in Practice

Sometimes when we have what we call a mental breakdown, it is actually the beginning of a process of change. In Michelle's case, this was surely true. The coping mechanisms we have learned in life actually hold us back from realizing a far more powerful thing that is within us—our innate mental health.

When Michelle started to mentally "break down," her defenses broke down as well; and so she became open to learning something new. This openness allowed her to listen and regain her inner wisdom. Learning to trust in yourself and the answers that lie inside is one of the most profound realizations we can receive in life. She learned to trust in the power of her wiser thought process, instead of the habit of worry, control, and analysis. By calming down, her natural state of mental health began to bubble to the surface. It was hard for her to accept at first, because a lifetime of control and struggle taught her to see what was happening as too simple to be real. For a workaholic like Michelle, doing nothing and trusting couldn't have been farther from her habitual way of coping with life.

When human beings begin to understand how to tap into this natural thought process, they see that life is easier and doesn't have to be an uphill battle. When we go with the flow of thought and wisdom, our lives fall into place with very little effort.

Trust in the power of the flow of thinking, and you can turn any crisis into a realization.

11

COMPUTER MELTDOWN

Frank has a major deadline coming up tomorrow. In the middle of writing the proposal that's due, his computer freezes up and he can't get it to re-boot. He tries to stay calm, but all he can hear inside his head is the voice of his irate boss saying, "That's no excuse! Why do you always screw up everything? I thought I could count on you."

With each passing thought Frank's fear and guilt grow. He tries to call the computer support line, but after about ten attempts to get through their automated phone system and another ten minutes of canned music on hold, his anger and frustration are starting to get out of control. By the time he actually hears a real human voice on the other end, he explodes.

"I'm sorry sir," says the impersonal voice on the line, "but we are backlogged with requests at this time and we won't be able to get to your request for at least a week."

"What do you mean 'a week'!" yells Frank. "I have a proposal that's due tomorrow." With that he slams down the receiver in frustration.

Now he's got to face his boss. "I know you aren't going to like this," he says sheepishly, "but my computer froze up and they can't fix it for another week. Can you get me access to another computer?"

"I can't believe this!" his boss roars. "I'm already backed up and my boss is going to jump down my throat, not yours."

Frank can only stand there, his face a picture of mute apology.

"I suppose I can find another computer somewhere," growls his boss, "but you'd better have this done by tomorrow!"

At 4:00 P.M. Frank finally gets the computer from his boss and sits down to work on the proposal. But now he can't think. His mind is racing, but he's so exhausted and emotionally drained that he's unable to focus on the task at hand. The proposal he writes is full of mistakes, and he knows it isn't his best work. By the time he gets home it's 11:00 P.M., and all he wants to do is sleep. But Frank's troubles aren't over yet. His wife is cold to him and says she doesn't understand why work always has to come first. They fight before going to bed, and Frank has a fitful night's sleep.

The Fork in the Road

Anyone who's ever depended on a computer as a tool for work can probably relate to this story. Who can blame poor Frank for having a meltdown? Wouldn't anyone react in the same way?

Not necessarily. Let's look at an alternate scenario for this situation. This time, instead of his mind churning out a full-blown "thought attack," he regains his bearings and handles this difficult situation with grace and a feeling of challenge and ease.

First of all, suppose that at the moment Frank thinks about his boss's reaction, he recognizes that his thinking is about to get out of control. He takes a deep breath and realizes he has a potentially challenging situation, but not a disaster. He imagines that his boss has had similar situations and will be understanding. Instead of calling the computer support service immediately, he walks out of his cubicle and asks Pam, a co-worker, for advice and support.

"Pam, you won't believe what happened to my computer just before tomorrow's deadline." He describes the problem to her. "Any suggestions?"

"Oh no! The same thing happened to me last month, and I couldn't figure it out. I guess you'll just have to call the computer service. But if I can help in any way, let me know."

Sound like a waste of Frank's time? Not at all. When Frank does call the computer service, he is calmer than he was in the first scenario and waits patiently to talk to a real person. It seems normal to him. When he does talk to the service person, he is calm and respectful instead of angry. He still is told it will be a week before they can help him, but instead of slamming down the receiver, he tries another tactic.

"I know you're backed up with other customers," says Frank sympathetically, "but my work is due tomorrow, and it's on that computer. I'm sure you've been in my shoes before. Is there anyway you could possibly speed up the repair call? I'd really appreciate your help."

"Most people in your situation aren't so understanding," says the service person. "I'll see what I can do about getting someone out there ASAP."

Frank is touched by the efforts of the service person and doesn't feel alone in his plight. They can't fix his computer, but they provide him with another one. He still has to rewrite the whole proposal, but because he has his emotional bearings, he rises to the occasion. His writing seems to come out effortlessly and in a quarter of the time it would have taken him in the earlier scenario.

His boss compliments him the next day on a job well done. "I can't believe you actually got the proposal done—and done well—under such pressure. How did you stay so calm?"

"I just didn't have a meltdown along with my computer," says Frank.

The Principle in Practice

How did Frank in the revised scenario stay so calm in the face of all those obstacles? He recognized that *his emotional experience was coming from his thinking*, and not the situation. This recognition empowered Frank to regain his emotional balance. When he regained his balance, he was able to calmly assess the situation and creatively see what his options were. In this healthy thought process, he *communicated more clearly*, saw the other person's point of view, and *didn't allow his fearful imagination to get out of control*. Most important to Frank, his thoughts *were flowing in an organized and intelligent manner*, which allowed him to complete the proposal.

What Frank tapped into is something that is only a thought away from all of us. We all have an *innate intelligence*, called *wisdom*, that allows us to rise above circumstances, and respond to life with common sense rather than react with anger, panic, and fear.

When life gets out of control, the first step is always to regain your bearings.

12

THE POWER OF PRESENCE IN RELATIONSHIPS

Allie attended a seminar I gave on how to prevent stress in the workplace. One of the exercises had to do with learning how to be present when listening. Although Allie had always thought of herself as a good listener, she was open to learning more. The day after she did the exercise, she shared this story.

"I was amazed at just how much I learned yesterday about listening. At first, it didn't really seem that different from what I'd already been doing. But when I arrived home last night and my two daughters wanted my attention, I had the chance to put what I'd learned to the test. I felt like I had a million things to do, but I decided to take your advice and really listen to my daughters with nothing else on my mind. When distracting thoughts would arise, I would just let them go and come back to being present with my daughters. Both of my daughters became very talkative and open about their day, sharing all the details enthusiastically. Their mood and mine lightened up. We didn't do anything special, but we had one of the most pleasant evenings I can remember—all because I truly listened.

"Then on the way to the workshop this morning, I was filled with feelings of gratitude for my daughters and my husband, and

I enjoyed the ride immensely. I can't believe that all I had to do to feel this way was to listen. It really makes a difference in relationships if we can stop our minds from wandering and truly be present with those we love. I can see the powerful effect listening has on feeling love for our families."

When Allie was finished, another workshop participant shared a similar story. Ken started by saying how skeptical he had been during the previous day's seminar.

"All I kept thinking was that I didn't have time to listen like that. I had too many other things to do in my life. All I could think about during the seminar was how I had to go to my office and pick up my e-mail, return phone calls, and then go home and do some chores.

"When I got home last night, I felt really wound up, but my boys wanted me to play a game of hide and go seek. It was a beautiful spring night, but I just felt overwhelmed. I gave into their pressure and began to play with them, but my mind was full of distractions about all the other things I had to do. This is usually how I feel when I'm with my boys—I'm torn between giving them my full attention and thinking about all the other things I *should* be doing instead."

The Fork in the Road

"Suddenly, I was struck by a thought. It occurred to me that as long as I was playing with the boys, why not enjoy it and let go of all my other obligations? It seemed foolish to be splitting my attention between my boys and my 'shoulds.' After all, it was only my thinking that was keeping me from being present in the moment. My preoccupation with other things was spoiling my enjoyment of spending an evening with my boys, and it was keeping them from really having their dad's attention. Sure, the thoughts about everything else I should be doing kept coming, but with less intensity.

"I had a delightful time playing with the boys and really got into it. At one point, I was hiding behind the mailbox in the front yard. My neighbors could see me, and it didn't even occur to me how silly I must have looked sprawled on the ground. I felt like a kid again. I even had time this morning to swing by the office on my way to the seminar and check my messages. A thought passed through my mind that perhaps I was feeling overly responsible about work—it really wasn't a big deal to wait a few more hours to get my e-mail. The time I have with my boys is so precious—and I don't want to miss a moment of it."

The Principle in Practice

Feeling intimate, close, or connected to others is the by-product of being fully present in the moment with a mind that is free of distractions. When we are with other people and our thoughts are flowing rather than preoccupied or obsessive, we feel calm, we feel present, and our listening is very deep.

Consider what you do when you use a walkie-talkie: You either listen or you press the button and talk. When you press the button, you no longer can hear the person on the other end. The same is true for listening in daily life; you are either listening or you mentally press the button and your mind is somewhere else. We have all had the experience of talking to people who are either not listening or momentarily off somewhere else. When that happens, our sense of connection to them is broken. We don't necessarily know why we feel disconnected, but we can always feel it, and we leave these kinds of interactions feeling empty and unsatisfied.

On the contrary, when we are with our child, partner, friend, or co-worker, and both of us are fully present, with minds free of distractions, both feel enriched by the experience. Couples feel love for each other, children feel cherished, and co-workers feel understood and appreciated. In our hectic world, taking the time and the presence of mind to be truly in the moment is one of the greatest gifts we can give to others as well as ourselves.

All it takes to be in the moment is simply to recognize when you are not and bring yourself back. Let the distracting thoughts go, even if they come back numerous times. Value *being present*, and you will have the kind of closeness in your relationships that you never dreamed you had the time for.

13

THE TRUE SOURCE OF GENIUS

As a child, Dave was considered a genius. He was in all the accelerated programs in school, and his parents and teachers expected that he would do great things someday, perhaps as a research scientist or theoretician.

When he was four years old, Dave's dad died suddenly, and his mom became very depressed. In fact, she remained depressed the whole time he was growing up. His older brothers and sisters followed their mother's example, and so the mood of his whole family was depressed. Although Dave was initially saddened by his father's death, he got over his grief in a reasonable amount of time, unlike the others.

"I was always trying to figure out why I felt like such an outsider," Dave told me. "I felt happy, and everyone around me was depressed. I tried to fit in with the others, but I never quite could. There was always this tension in me—trying to hold back my natural feelings of happiness and questioning why I felt so good when others around me felt so sad. I began to think that they were normal and that there must be something the matter with me.

"I tested for Mensa, and they found I had a genius IQ. After high school, I was accepted to study to be a neurophysiologist researcher at the university. Everyone constantly reminded me of the gift I had been given and my responsibility to do something with it. Although school came easily for me, I was always afraid I wouldn't live up to the expectations of others, so rather than study, I would goof off. I became extremely anxious about becoming a failure; I was so afraid of failure that I wouldn't even try."

Then, at age nineteen, Dave discovered alcohol.

"It was fantastic," he said. "I'd never felt that good in my life. As soon as I had a drink, all the feelings of self-doubt and tension immediately disappeared."

Dave continued on in school, drinking more frequently and devoting less time to his studies. Despite his drinking, he did manage to graduate after six years and landed a great job. But he had stopped growing psychologically. Though he was told he was brilliant by his professors, he was constantly disappointed in his own performance. He shared with me that he suffered from what he called "must-urbation"—"I must do this, I must do that." He was very critical of himself and perfectionistic. During his twenties, he led a double life. He would work at respectable jobs during the day and hang out in seedy bars at night. No one in his family knew he drank that much. He was aware he was killing himself, but didn't seem to care.

Dave eventually ended up in treatment for alcoholism. As soon as he got out of treatment, he immediately started drinking again, even though he went to AA. His situation deteriorated to the point where he could only hold temporary jobs, and he struggled from paycheck to paycheck. Then one day a friend offered him some crack cocaine.

"I was hooked in fifteen seconds," he said. "I felt omnipotent. This started my career as a drug addict. You might wonder how I could have become such a loser, when I had so much potential. But my self-esteem was at rock bottom. I lost my apartment and got several DWIs. Finally, I decided to take a lethal dose of Dilantin, actually six times the lethal dose. The next day, to my disappointment, I woke up, still alive, though I could hardly walk. I stumbled to a pay phone and called my mom. She came immediately and put me into the hospital to detox from the Dilantin. She never gave up on me. And so I went through commitment proceedings and gladly signed the petition that forced me to get help."

The Fork in the Road

"After my detox, something began to change. I found out about this new treatment center with an approach based on the three principles of mental health, and I began to feel hope because it wasn't a traditional disease-based program. Instead, it focused on bringing out the health in addicts. I started to think that perhaps there was a force at work to keep me alive.

"For the first time, I entered treatment with a positive attitude; I was open to anything they had to offer. When I met one of the other 'students,' as they called the patients there, he began to tell me about the program. He told me things I had wanted to learn about all my life—how my mind worked. It was so simple and commonsense, and I felt more powerful feelings than I had that first time I used alcohol or crack. I was filled with hope and a serenity I had never experienced before, and no drug had ever come close to giving me that. The turning point for me was realizing that those positive feelings were natural, that they were always inside of me, just waiting to come out.

"Through the treatment process, I discovered that I already had innate mental health within me. For the first time, I realized that I didn't have to figure it all out analytically, even if I did have a genius IQ. I felt so relieved that I could trust in something deep inside me that would guide me in life. Finally, I learned to truly trust myself. In the past, I used to have tons of thoughts. But now I learned to trust my wisdom instead of my analytical thinking. I could let go of the negative thoughts and trust that the wisdom would come.

"When I let everything go and just trusted that the wisdom would reveal itself, that was the real fork in the road for me. As long as I was in the moment, I was fine. My wisdom was there, and I knew I would be all right. When I discovered the power of living in the moment, everything in my life fell into place. I was able to stabilize my moods. And when I even think about using cocaine, I know it is only a thought, and it simply goes away. The other day, I got a migraine and realized that my mind had just gotten superbusy. All I had to do was realize this, and my headache went away too. It was such a relief to discover that I didn't have to think all the time, that self-analysis and a highly developed intellect weren't the answer to my problems—wisdom is far more powerful."

The Principle in Practice

If happiness came from intellectual intelligence, Dave would never have become an addict. Prior to learning about the principles of his mind, he was always thinking, always trying to figure himself and his problems out. When Dave discovered the intelligence that is greater than intellect—wisdom—he could relax and trust that it would guide him in life. When he calmed down and had faith in this wisdom, he *knew* he would be all right.

Serenity is both the goal and the means to the goal of recovery from addiction or from any psychological problem. To achieve serenity, we must let go of our analytical thought process and trust in our wisdom. When we live in the moment, serenity is always there. It is only when we project into the future or the past that we lose our serenity and become anxious, depressed, or experience some other negative emotion. Dave discovered the power of the moment to bring him back to sanity and to his true genius.

You don't have to be a genius to have wisdom; it is available to all of us. When your mind spins out of control, let go and trust that the wisdom within you will guide you back to happiness. Wisdom is true genius.

14

FALLING IN LOVE WITH YOUR JOB

When Jean started her business as a medical-practice consultant in 1985, she was full of enthusiasm and energy. She wanted to help physicians deal with the difficult changes in the insurance industry that affected reimbursement for their services. She knocked herself out trying to respond to their every need—crisis calls at all hours of the day and night, early morning meetings before they did surgery, late night calls if they were in a panic. She tried so hard to please them, to make everyone happy.

Within six to eight months, Jean was totally exhausted and burned out. She would stand in the shower each morning for ten to fifteen minutes, banging her head against the tiles, sobbing, "I'm so unhappy. I hate my job!"

Jean also felt trapped. She had borrowed a great deal of money to start the business, and she had many loyal and good employees, leases, and other obligations. Then one day a client of hers recommended a prospective client, a psychiatrist who had a very successful practice, but had no time to run the business end of it. "No way!" Jean said. "I don't want any more clients. I'm already overloaded."

Her client pleaded with her and said, "But Jean, this is a good friend of mine. Please do this for me as a personal favor."

Reluctantly, Jean agreed to have lunch with her client and his friend to consider it. As she looked at her client, who was showing signs of stress, and reflected on her own burned-out state, she noticed that the new client prospect, a Dr. Jones, was the only happy one at lunch that day. Intrigued, she decided to meet with him again. She learned that he did indeed have a very successful psychiatric practice in terms of number of patients; however, his practice was inefficient from a business standpoint. She also knew she could help him.

Dr. Jones suggested that she sit in on his weekly aftercare group to get a feel for what he was doing, since it was quite different from most psychiatric practices. He was using a new model of psychiatry based on teaching health rather than treating disease. Although his patients were only in therapy on a short-term basis, he seemed to have had remarkable success with them, especially in comparison to his colleagues who treated maybe a couple hundred clients for many years each.

Jean came to the evening group feeling very skeptical about how this doctor could possibly give successful treatment, and feeling superior to these "mental patients." What could she possibly learn from those people? After all, wasn't it true that she never had mental problems and had been a very successful businesswoman most of her life?

As Jean half listened to the group, she began to observe that the patients appeared to be totally normal, despite the fact that they had been previously diagnosed with serious mental illnesses such as bipolar manic depression, posttraumatic stress disorder, and depression. Yet now they were living perfectly normal and happy lives. They were just grateful that they had learned how to use their minds in a healthy way. Jean was struck by how calm they were about it, neither excited nor evangelical like so many others she had witnessed. *These people are happier than I am*, she realized. *But they used to be mentally ill. How could this be?* For the first time she began to feel hope that she could be happy too.

After a few weeks of going to Dr. Jones's group, Jean noticed that she was taking weekends off to be with her daughter, a new phenomenon for her. One Monday morning when she was in the shower, she realized she wasn't crying, as was her habit. She realized that she was actually looking forward to going to work. Nothing had changed in her circumstances—the doctors were just as demanding; the debts were the same. Self-righteously she rationalized that maybe she was finally getting a handle on this business.

The Fork in the Road

That day she went to work and immediately received a call from one of her physician clients. "I need to talk to you immediately," he said anxiously. "Something terrible has come up."

Although she knew she had time right then, Jean replied, "I can't see you right now, but call me after lunch and we'll get together." This was strange behavior for her as well, because she usually jumped every time a doctor needed anything.

Later he called back and apologetically said, "I'm sorry I reacted so poorly this morning. I was in a terribly frantic place and treated you in a demanding way. Whatever it was has taken care of itself. Thanks for all your help."

This really puzzled her. She had *never* had a doctor thank her, let alone apologize. That week when she went to Dr. Jones's group, several people asked her if she had been on vacation, because she looked so relaxed and rested.

"As a matter of fact, I worked all week," she replied. That night she listened better, and something Dr. Jones said hit her with tremendous force.

"The problem with most people," he said, "is that they live as if someone were writing them hate mail. They go to the mail box, read it, and then feel upset by what they read, not recognizing their *own* handwriting." She realized that night that *she* was the author of her own hate mail—other people's ingratitude, their demands that she make them happy. All this was her own thinking! She realized that she had created this world of extreme pressure and frenzy. She was always trying to prove herself and was a victim of her own success at doing it.

Jean left the meeting that evening elated. When she walked through the door of her house that night, she felt as if she were beginning her life anew.

Jean continued her consulting business for a few years, but her focus shifted from fixing all the doctors' problems to empowering them to fix their own. She used to uncover the problems, prioritize the seriousness of each, then tell them what to do. Just when she fixed one problem, another one would jump out like a jack-in-the-box. She didn't have enough hands to keep all the jack-in-the-boxes under control. Consequently, she and her clients lived in a state of frenzy and reaction.

Now Jean teaches her clients that problems are thoughts. She neutralizes their thinking by showing them the power of thought to create their experiences. She encourages them to see their work and difficulties with a fresh outlook, a healthy state of mind. This allows them to come up with their own, creative solutions. Jean loves her job now and has even expanded her work to serve on the faculty at a university and does consulting for other kinds of businesses as well.

The Principle in Practice

Before she met Dr. Jones, Jean was a burned-out, cynical person. She had very little hope of ever changing her life and felt destined to a life of disappointment and endless efforts to please others. When she saw the changes that others had made in achieving a happier life, people who'd had problems more serious than hers, it gave her hope—her mind opened to new possibilities.

Jean discovered the power of changing her thoughts, rather than changing her job, to make her happy. When people become stressed and burned out over time, they need to rediscover hope in order to open their minds to change.

Hope opens the door to the possibility of change.

15

ANGER: IT'S AN INSIDE JOB

"You jerk! I can't believe how inconsiderate drivers are these days." Such are the angry musings of Hugh on his morning commute to his downtown job in a high-tech firm. It's a typical Monday morning as Hugh barrels down the highway. He aggressively attempts to edge his way into the left passing lane, but the woman in the blue Buick stubbornly hems him in. "Another idiot!" he fumes, his blood pressure rising as he grips the steering wheel tighter. "The road is full of them this morning."

The driver of the hated Buick avoids his cold, angry stare as she applies more eyeliner to her left eye while glancing at the visor mirror. Not to be ignored, Hugh lays on his horn as he raises his other fist in defiant protest. She replies with a mouthed "Up yours, Buddy!"—and the war is on.

He mutters to himself, "Okay, lady, you wanna play chicken, you've met your match," as he sees his moment of opportunity and kicks it into fourth gear, moving over into the right lane to pass—and barely missing the red Ford Bronco on his right.

"Gotcha," he thinks smugly to himself. What he doesn't realize is that, although he may have won this small battle, he is loosing the war. Once again, he arrives at work frazzled and angry, which sets a negative tone for the rest of his day.

The daily battle through rush-hour traffic and scores of

inconsiderate, hurried drivers has become a road well traveled for many modern urban dwellers. Traffic—the ultimate form of powerlessness—leaves many of us stressed, frustrated, and enraged. There is even a term for this syndrome: *road rage*. There are books, articles, radio talk shows, and therapy programs that deal with this growing phenomenon. How do we avoid the stress and strain of the daily commute and move into the "sane lane"?

If we could look into Hugh's head we would see a mind full of judgment, impatience, and intolerance. To him, other drivers look like the "enemy" or, at the very least, people who are out to make his day more difficult. In Hugh's mind, the drive to work is a war zone. He takes every intentional as well as unintentional discourtesy personally, and takes it upon himself to administer the proper punishment.

Hugh is unaware that his thinking and mind-set have anything to do with his experience of driving to work. He sees his experience as a natural reaction to what occurs "out there," without any awareness that his psychological life is created from the "inside out." How would traffic look to Hugh if he realized where his experience was *really* coming from?

In an ideal world, traffic would always flow smoothly, drivers would all be courteous, and there would never be any delays due to weather or accidents. Unfortunately, this "ideal world" only exists in the expectations of human beings. Accidents do happen, the weather changes on a daily basis, and people, being human, are not always in a positive, courteous mood. But, as a driver on the road, we always have a choice as to how we will respond to the normal variances of life—we are always at the psychological "fork in the road." Once we realize we have a choice, we can begin to make that choice: to go in the direction of a negative emotion or seek out harmony and enjoyment in our everyday lives.

The Fork in the Road

What if Hugh finds himself in the same situation with the woman in the blue Buick, but this time he realizes where his rage is coming from? He feels impatient that traffic is not moving very fast this morning. If it doesn't speed up, he worries, he'll probably be late for work. As his tension begins to build, he notices his shoulder muscles tightening and his irritation growing.

Hugh realizes that his feelings of impatience, irritation, and discomfort are signals that he is caught up in his thinking and needs to make an attitude adjustment. Like the bumps that wake us up to the fact that we are drifting over into another lane, his emotions wake him up to the realization that his thinking is drifting over into the unhealthy zone. If he keeps it up, he is headed for road rage.

Simply by recognizing the undesirable course of his thinking, Hugh is mentally able to shift gears. Instead of seeing the woman in the blue Buick as the enemy, he is amused that she is doing two things at once. He realizes that she is too preoccupied with her eyeliner to notice he wants to change lanes. So he slows down and lets her move ahead, giving him plenty of room to pass her. He even has to chuckle at himself for almost making ten seconds of wasted time more important than enjoying the ride and arriving at work relaxed and in a good mood. Once again, Hugh realizes that circumstances don't dictate his moods. He realizes the power of his own thinking to create a life of rage or a life of calm.

The Principle in Practice

By recognizing his uncomfortable emotions, Hugh woke up to the fact that he was thinking in a way that was creating his road rage. Just as physical pain is there to alert us to the fact that we are moving out of physical balance and health, uncomfortable emotions warn us when our psychological health is becoming unbalanced. Once we understand the role of *emotions as a signal* to alert us to the quality of our thinking, we can change the daily commute from a time of tension and stress to a period of relaxed quiet time.

Hugh also realized a second important concept in regaining control of his life—the source of his psychological experience is not external circumstances, but rather how he sees and responds to those circumstances in his own mind. Whatever we are doing in the moment, we are always thinking simultaneously. Our experience, feelings, perceptions, and sensations are all coming from what is on our mind, not the external situation. *Our experience of life is created from the inside out, not the outside in.*

Learn to listen to your emotions as an inner signal about the quality of your thinking.

16

THE OAK TREE IN THE ACORN

Before Officer Mike came to our neighborhood to be our beat cop, he was the personification of "Miami Vice"—a rough and tough cop from Florida who saw police work as a war of good versus evil. He'd had more complaints from his local Florida community for being physically abusive than any other cop.

Then Mike went through a change. He realized that there was a kinder way to help others, that there was more power in fighting crime with understanding than with force. He became very committed to community policing, and at that point he moved to St. Paul and began to work in our neighborhood. Full of enthusiasm and energy, Mike drew people to him. He was the "pied piper" of Fairway Avenue; all the kids would follow him everywhere. Our neighborhood began to be a safer, friendlier, and more positive place, and Mike was a big part of that. But for him, there were still more changes to come.

Inside, Mike was very busy-minded. He always had too many cases to handle, would work overtime without pay, and saw himself as striving single-handedly to eliminate crime from the neighborhood. Often, he felt alone in his plight and would resent the other officers for not doing their part. He took every crime personally, as though he had somehow failed in his efforts. His head was filled with thoughts: positive ideas for improving the neighborhood, many of which he was never able to complete, and many negative ones of feeling overwhelmed, insecure, and anxious.

Late at night, Mike would arrive home exhausted and plop in front of the TV to find some peace. This was the only time his mind would slow down. His wife, Cindy, would complain to him that he was never there, and his kids would compete for his attention, often acting out. But Mike only saw his wife as a nag who didn't understand the pressures on him, and his kids as demanding pests who wouldn't leave him alone. Unable to turn off his thoughts, Mike couldn't sleep at night. He was exhausted and would often become ill—his one escape from the pressures of being a supercop.

When I met Mike, he was attending a talk I was giving on crime prevention and how to deal with being a victim of crime, one of a series of lectures for the local area. He later shared with me, "I thought you were a total simpleton who didn't have a clue about the complexities of what a cop had to deal with."

After avoiding me for the next six months, one day he came to my office, and we just talked. He listened for the first time and really was affected by what I had to say about thought. And, for the first time, he realized that he was always thinking.

The Fork in the Road

Mike later said to me, "I had no idea how anxious I was all the time. I had never noticed how I was feeling and just how busy my mind was. Gradually, I began to just notice my thinking and not take every thought seriously. And I calmed down somewhat. I quit analyzing every thought that occurred to me—all my worries about the other officers and my supervisors, my speeded-up thoughts about positive ideas, my feelings of responsibility for changing the world. I began to experience the health in myself and see it in other people, and that allowed me to calm down inside."

To share with me how this realization changed all aspects of his life, he told me about something that happened one night when he got home from work. As usual, it was late, and Mike was greeted at the door by Cindy's litany of complaints about him and the kids' behavior. "Why can't you help me out with these kids?" she said angrily. "Your son and daughter have been at each other all day, fighting. You spend all your time helping other parents' kids and ignore your own."

Mike could feel himself tense up and become defensive. *She doesn't understand how important my work is. You'd think your own wife would understand. If she only knew what I have to go through on the street.* These thoughts flew through his mind, but he recognized that they were only thoughts, so he kept his mouth shut and decided to just listen to his wife.

As he continued to listen, he began to stop seeing Cindy as a nag and instead perceived her as a person who missed him and loved him and was feeling insecure. He said to her, "I can't believe I haven't heard you before. I'm sorry, it must be really hard on you to be all alone with the kids all day and feel so unsupported." He stepped forward and gave her a long hug. She relaxed in his arms, and for the first time in a long time they felt close again.

Mike told me about how this one moment transformed his home life. "I began to really listen to my wife and kids," he said, "and when I did, they calmed down too. As my mind began to clear of all the burden I had felt on me to change the world, I began to see the beauty in my family and their need for me. I felt like a burden was lifted off my shoulders. I began to see the oak tree in the acorn. I started seeing the health in the people on the streets—the gang members, the homeless, my fellow police officers. I started listening to the feelings I was having, instead of focusing on the content of my thinking. It became natural for me to listen more and just *be* with people."

No longer seeing everything in terms of good and evil, Mike started seeing the common humanity we all share—the true innocence of people who are caught up in their thinking and then act on those thoughts, just as he had done. He saw the kids who carried guns as very frightened children who were trying to protect themselves. As he began to understand them and quit judging them, they felt his respect and would listen to him. He began to teach them about their thinking and how fear was created with thought.

Mike's impact on the neighborhood and within the gangs was so powerful that the media began to do stories on the transformation happening in our neighborhood and the efforts of this one dedicated police officer. Many of his co-workers began to attend training sessions and, as a result, a change is under way in our police department.

A few months ago, Mike took on a new position and left the neighborhood. Now he is the coordinator of gun violence prevention for the whole county, and he is taking the simple message of how thought creates all our realities to numerous groups in the community. He is even beginning to teach this approach to officers in other parts of the country. Although Mike is still the loving and compassionate officer he was before he understood the power of thought, he is now calm inside. People who know Mike are amazed at the change that has taken place within him—some are drawn to him, and others are threatened that he isn't as anxious as they are.

The Principle in Practice

When we see life as a battle between good and evil, we automatically create an "us against them" mentality. With this mindset comes fear—fear of the power others can have over us. We see others as needing to change, and we also see it as our job to change or control them for their own good.

When our mind is full of judgmental thought—this person or behavior is good, that one is bad—we create the illusion of separateness in humanity and avoid seeing our commonality as human beings. As Mike began to recognize his feelings and his thinking, he relaxed and slowed down. Immediately, he became a better listener with his wife and family. The burden of changing them, saving them, and convincing them was transformed into loving them and listening deeply. As a result, they have a closeness I have rarely seen in a family.

Instead of seeing what other people were lacking, Mike began to see the potential that lies in all people. He felt love for them. He began, in his own words, to "see the oak tree in the acorn." The impact of seeing past people's exterior behavior to the common humanity we all share has not only transformed Mike, but also has affected the hundreds of people whose lives he touches.

When I asked Mike what his vision for the future of policing was, he shared this with me: "Policing as we know it will be obsolete in ten years. Our work now is based on changing criminals through fear of punishment. I have never really seen this work. Instead, it has created a division in our society between the criminal element and those we think of as the 'good people.' This division is continuing to grow and expand as our prisons fill up more each day. Once we begin to see the common humanity in everyone, we will use love and understanding to help others rather than wage a war on crime. That metaphor itself is all wrong. As long as we see it as a 'war,' it will get worse and alienate people further. But if we can teach people to use the power of their minds to create their lives, they will regain hope and the power to change. Then we will begin to eliminate crime from our society."

Remember, the oak tree is already in the acorn; it only needs love and understanding to bring it out.

17

FAT-FREE THINKING

Fran always thought of herself as the weak one in her family. In fact, her dad used to call her "Runt." When she was six, she asked her older brother what "runt" meant. He said, "The runt of the litter is the smallest and the weakest and it usually dies." His words not only frightened her, they also had a huge impact on her self-image. And so Fran began to think of herself as weak and incompetent and played out that role in her family and later in life.

As a coping mechanism for dealing with her feelings of being weak and incompetent, Fran began to find comfort in food. She discovered in food a form of relief that would momentarily make her feel strong and definitely not like a runt! Consequently, she became very obese and for most of her life was sixty or seventy pounds overweight.

As an adult, Fran continued this habit, even though she did survive and was no longer weak. She was very anxious and used food to calm herself down. A binge eater and a yo-yo dieter, in any six-month period she would lose up to fifty pounds and then put it right back on. Fran's entire self-image was connected to food. If she was eating properly, she was an okay person; if she wasn't, she was incompetent and a bad person. But no matter how much she weighed, she was always thinking about food. *I should eat this. I shouldn't eat that. I want that cookie. That's bad for you. You're a fat slob. You're doing better now. Just a few less pounds and you'll really be doing great. But will it last? You're never able to make it last.* Fran's mind never seemed to be free of food-related thoughts.

The Fork in the Road

Fran told me what helped her to change her life. "After I learned about how my mind works," she said, "I began to calm down immediately. I was far less anxious, but I also began to eat differently. I now see that I am already an okay person, no matter what my weight; I don't have to lose weight to be a good person. Knowing that I'm already okay and knowing how to calm my mind helped me begin to even out emotionally and get off the weight-loss roller coaster. Since I'd always ultimately end up at the same weight after six months of yo-yo dieting, why not just accept myself as I am? At least I would be happy during those six months instead of miserable.

"The irony of it is that when I quit making food such an issue in my head, I began to gradually lose weight anyway. The next six months after realizing this, without really dieting, I lost forty pounds and have kept it off since then."

Over the past year and a half, I saw Fran every month or so. And every time I saw her, she looked like a different person. Her face would get softer, her posture would be more confident, and she looked healthier and happier. And she continued to lose layer after layer of extra weight.

Fran explained to me how it felt for her to undergo this transformation. She said, "I felt like the strong person who was buried under the layers of anxiety and insecurity began to emerge. For the first time since I was very young, I was being my *true self.* As the layers of insecure thought fell off, so did the weight. It was very natural; there was no struggle involved. I simply began to recognize my feelings and sensations. When I was hungry, I would listen to what my body was hungry for, rather than listen to what my brain told me or eat the first thing I saw in the refrigerator. When I was full, I would also listen to that and stop eating. In the past, I had buried my body's natural mechanisms for weight management by not listening to the inner signals of hunger and fullness. I don't diet now, but I do eat healthier. I exercise because it feels good, not to lose weight. And I feel great!"

Fran looks great too. She radiates a true feeling of serenity and confidence, because she loves herself and her body. Because she is more open to others, she has more satisfying relationships as well. Fran is an excellent example of the power of listening deeply to the wisdom that is always guiding us toward health.

The Principle in Practice

People spend billions of dollars per year on weight-loss programs, diet foods, exercise equipment, and health-club memberships all with one goal in mind—*to lose weight*. We direct much of our attention to what we eat, how healthy we are, and how to improve our bodies. Yet many of us are still obese and fight continuously with our weight. And despite the fact that we have the most expensive health-care system in the world, behavioral causes of illness such as overeating, smoking, and drinking give us very high rates of heart disease, diabetes, and other lifestyle ailments. Why is this?

Fran is a perfect example of how a person can lose touch with the true source of feeling good. Instead of finding a healthy state of mind and then letting her inner wisdom guide her back to a healthy body and psyche, she was trying to *feel good* first through food, then through losing weight. By learning the principles of psychological health, Fran was able to find peace of mind. It was only when her mind was peaceful that she was able to listen to the *natural* homeostatic mechanisms (self-correcting processes that automatically signal our body) that are built into us for weight control, exercise, and health.

The human body is a miraculous demonstration of the power of health; it can heal a broken bone, grow skin over a cut, or recover from an injury. It does this with very little assistance from us. The same is true for the miraculous power of our minds. When we get our unhealthy thought processes out of the way, our inner psychological health heals the past, overcomes harmful habits of thinking, and assists us in listening to an enormously intelligent guidance system.

Learn to be happy and to accept yourself no matter what your body looks like. When you do, your perfect state of health—physical and psychological—will emerge.

18

PUTTING JOY BACK INTO WORK

As a child growing up on the tree nursery my parents owned in the Midwest, I learned the values of hard work, keeping busy, and getting things accomplished. All of this became tied to my feelings of self-worth and my day-to-day moods. If I didn't get much done, I felt lazy, depressed, and guilty. If I accomplished many things, I was in a good mood. But no matter how many tasks I completed, I felt driven by a sense of urgency to get even more done and would experience the actual work as stressful, exhausting, and unpleasant.

The upside of this work ethic was that, as an adult, I continued to be extremely productive. My wife was amazed at just how much I could do in a day. I could clean the whole house in a couple of hours, write a business proposal in a morning, and plant all the flowers in a single day. The downside was that I never really had fun during the process, and I was unable to just sit down and relax without feeling guilty or anxious. I would think of one more thing to do and then still one more, and this would go on until the day was through. For me, the greatest challenge in life was *sitting still*.

When I learned how my mind works, the first thing I had to

learn how to do was *nothing*. My mind was so busy that to sit by the beach and just soak it all in—see the beauty, smell the scents, hear the ocean—was a whole new concept. I was so afraid that if I wasn't speeded up, I would become lazy and not accomplish anything in life. This thought would make me depressed, so I just stayed on the treadmill of making lists and checking off items as I completed them, and only then would I be able to feel good about myself.

The Fork in the Road

I remember clearly the day I finally realized how to sit still and not feel bad about it. I was in Florida for my three-week internship on a new development in psychology. For me to take three weeks off work was in itself a major accomplishment and caused me much anxiety.

The first two weeks of the program, I was full of questions. I desperately wanted to learn this new approach and understand how to help others through counseling. Unfortunately, I hadn't yet gained an understanding of the approach for myself. The leaders of the program kept telling me to just listen and to relax. "Take the afternoon off and go to the beach," they'd say. "Try to slow down, Joe. You'll be amazed at how much you'll learn."

Yeah, right! They want me to take the afternoon off so they can take it off, I mused suspiciously. *What a scam!* The thought of doing nothing for an afternoon but sit on a beach sent chills of apprehension through my body. I had never just done nothing. I argued that I only had a week left of the program and it didn't seem like I was getting anywhere, so shouldn't I try harder instead of going to the beach?

But then again, I thought, *I have nothing to lose. I could use the vacation, and maybe they're right. Maybe I do need to learn how to slow down.*

The next day I sat on the beach, went for a stroll with my wife, and actually enjoyed myself. But after an hour or two of this, the anxiety began to build up again. Nevertheless, I was determined to tough it out and sit there on the beach, no matter how uncomfortable I got. I realized that I needed to take a leap of faith and just trust that I would discover what my teachers were talking about.

That night, I went to bed and fell into a deep sleep. At 3:00 A.M., I awoke from a dream. Suddenly it hit me, and everything I'd been learning made sense. "Honey, wake up," I said as I shook my wife. "I got it! I think I understand what they've been telling me." My head filled with a flood of insights. For the first time in my life, I realized the power of letting go and not trying to figure out an idea. Because my mind was relaxed while sleeping, the understanding just came to me effortlessly. It almost seemed too easy and too good to be true.

When I returned to Minnesota and got back into my routine, the insight I had had that night kept recurring to me. One Saturday I was rushing through my to-do list when I *realized* I was rushing. I stopped and took a break and got my bearings back. *Maybe I should try this and see if it's possible for me to actually get anything done if my mind is relaxed instead of being in my usual stressed-out state.*

I proceeded through my day, but for the first time I made the conscious choice to stop myself periodically to take breaks and relax my mind. Sure enough, by the end of the day, I had gotten more done than I had actually planned to do. And what was most surprising is that I had enjoyed myself the entire day—both work times and break times—and didn't feel exhausted at the end of it.

The fork in the road for me was learning that it was perfectly okay to be happy no matter what—even if I didn't get anything done. I also realized that I didn't have to feel guilty in order to get motivated. Because I learned to have faith in the existence of my deeper intelligence or wisdom—which would guide me when my mind was relaxed—I realized that as necessary tasks came to mind, I no longer needed to go through a long, drawn-out, inner dialogue in which I'd browbeat myself into getting motivated. Now my motto was the same as Nike's: *Just Do It!*

To my amazement, I discovered that I could actually be very successful, get things done at work and home, and still have times in which I did nothing but relax, listen to the birds, watch the sunset, hold hands with my wife, and not feel guilty or anxious. In fact, I learned how to really enjoy life in the moment. The added bonus was that when I was actively *doing*—raking the yard, cleaning the house, paying bills, or organizing a room—I was getting things done effortlessly and enjoying those moments as well. Consequently, the lines between work and fun began to blur.

The Principle in Practice

One very important thing I realized is that human beings are not intrinsically lazy, but are rather *intrinsically motivated. Being intrinsically motivated means that motivation is natural.* To observe this principle in action, just watch a child at play for hours building a sandcastle, making a fort, or learning to walk and talk, and you will witness intrinsic motivation. To us, it might appear as if the child is only playing—*child's play,* we call it. However, children at play are actually learning very important skills—language, motor skills, coordination, teamwork, imagination, and creativity—some of the most important skills we can acquire, all in the name of fun.

It is only when we begin to think of learning as "hard work" that we take the fun out of it and put the drudgery into it. Because we don't trust our own natural motivation, we create a state of internal mental slavery and whip ourselves into action through negative self-talk like, "You are so lazy today. Get up and get going." Many of us listen to motivational tapes, make lists, and develop elaborate goal- and time-management systems mostly because we assume that without those tools we wouldn't get anything done.

When you begin to have faith in your innate mental health and your inner wisdom, you will be surprised to find that you are more motivated to do things, and that while you are in the process, it will be more enjoyable. You may even find yourself "whistling while you work."

Tap into your natural motivation, and you'll put the fun back into work.

19

"DADDY, MY THOUGHTS CHANGED!"

Kevin, a good friend of mine, was in the kitchen cooking dinner when his son, Cory, ran in the house yelling and seething with anger, stomping each step of the way. "I hate Molly. I'm going to kill Molly!"

"Why don't you come over here and sit down," Kevin said, indicating the kitchen stool, "and tell me about it?"

Cory sat down on the stool looking very angry. His fists were clenched, his face was a bluish red, his arms were crossed, his lower lip protruded in a pout, and his jaw was firmly set. Kevin knew there was no way his son was going to let go of this one right away, so instead of attempting to talk Cory out of his mood, he turned to stir the sauce on the stove.

The Fork in the Road

Not more than a half minute later, Cory broke the silence by asking in a softly curious tone, "Are you going to add all of the milk to the sauce?"

Kevin could hardly respond; he was shocked that his son's mood had changed so dramatically. One minute he'd been full of hatred and rage at his sister; the next, it was as if it had never happened or was somewhere in the distant past. Kevin observed that Cory's eyes were clear, his fists were relaxed, his posture was curious and positive, and his skin had returned to a normal color.

Puzzled, Kevin smiled and said, "Cory, you were angry at Molly a few minutes ago, weren't you?"

"Ya," Cory replied in an equally puzzled manner, as if wondering what that had to do with pouring more milk into the sauce.

"But you're not angry now, right?" Kevin probed.

"Right."

"What happened?"

Cory pondered this for a moment. Then he got this twinkle in his eyes and said, "My thoughts changed!"

When Kevin shared this story with me, he said, "I was so pleased and delighted that I walked over and gave Cory a warm hug. It was such a perfect example of how our thoughts directly create our experience of life. Then he and I talked about how nothing had really changed. His sister was still out in the backyard. The circumstances had not changed, yet he had completely changed—from being a boy filled with rage to a picture of relaxed curiosity in ten or fifteen seconds."

From being around Kevin and his family, I've observed that Cory and his sister, Molly, normally get along wonderfully for siblings who are six and nine years old. But like any brother and sister, they have their difficult moments. The beautiful thing is that not only do they get over things quickly, as many kids do, but that they also realize *how* they get over them. Kevin was delighted more with Cory's insight that his thoughts had changed than with the fact that his mood had shifted.

Kevin and his wife, Linda, had begun learning about the principles of healthy psychological functioning before their kids were born. I asked Kevin if their understanding of the principles had made a difference in parenting their children.

"Our awareness of these principles has slowly grown over the years," he told me. "We're still very human and have our moments when we get into a spat or one of us will be in a low mood, but overall I believe we show our kids what it's like to be in a healthy state of mind. Even if there is tension between us, we'll talk to the kids about it when it's over and explain to them about low moods and how our thinking changes. I think the kids aren't frightened of our low moods, because they know that the negativity is temporary; they know we'll return to our health. It's not that we lecture our kids about the principles, but we do find teachable moments. We treat their difficult moments or ours as opportunities to teach them about thinking and moods and how we create our experience.

"This understanding has given me confidence in myself and in life. I don't overreact to times when I'm off or the kids are off. I don't blow things out of proportion. For example, if I get angry at the kids and send them off to their room and later realize it had nothing to do with the kids, that it was just my overreaction to my low mood, I don't have to think, 'Oh God, have I ruined my kids? Will they hate me for this when they're fifteen?' I can put it into perspective and see that I'm just human and not worry about those things. They know this too and are growing up with a feeling of deep security, a realization that people do screw up because they're human, but that they'll be okay again later. They know that on the other side of the anger is compassion and love, that our health will return. There is tremendous comfort for them in that. They don't have to panic if their parents are having an argument, because they know we'll forgive each other and it will be over. They know if they get angry or upset, their state of balance will return. They don't have to be perfect."

What Kevin said reminded me of something that happened during their recent visit. I had told Cory not to turn off my computer when he was finished playing video games, but to ask me to do it. (He had pulled the plug out on a previous occasion.) However, once more, he turned it off, despite my warnings. When I confronted him about this, he wasn't afraid, even though I was this big adult and he knew he had made a serious mistake. He apologized to me, but he knew it would be okay. He wasn't scared to death. He knows we can forgive each other.

Kevin reflected, "I think the biggest way this understanding has helped me is in my ability to forgive myself as a parent. My imperfection as a parent comes out daily. In a life where grace exists always, forgiveness is always a moment away. That allows me to go through life with ease and not get caught up in negative thoughts like worry, guilt, and resentment."

The Principle in Practice

One of the underlying concepts of psychological well-being is that children are born with mental health. Cory and Molly are wonderful examples of this principle. Kids, like adults, have their ups and downs, their high and low moods. However, children have not yet learned to hang onto thoughts of the past, and so they usually do not process the experience for long periods of time. Cory was in a rage at his sister one moment and delightfully curious about why his dad was pouring the milk in the sauce the next. This is an example of how *moods are created moment to moment through our thinking.* When our thoughts change, our moods change. It is only when we persist in thinking the same type of thoughts that our moods persist.

Mental health does not mean that we are perfect—that we are always even-keeled and positive. *Mental health is being in a flowing motion of thought.* Letting ourselves be human is the greatest gift we can give ourselves and our children. It is totally normal and healthy to go up and down all day long, moment to moment. The beauty is to see that *we can and will change.* When we know this, we feel secure and hopeful. We know that thoughts are like weather; just wait a minute and they'll change.

As we understand the normalcy of mood changes and how life can look so different depending on whether we're in a low or a high mood, we can forgive ourselves and each other. Forgiveness is the dustpan of human relationships. We will forever be making mistakes and taking things personally, but forgiveness allows us to go on to the next moment—fresh and clear of the past. That's why Cory was puzzled when his father referred to his previous moment's anger. He was already in a new reality, a new moment.

As parents, the best we can give our children is the gift of our own mental health. We teach by the way we live, not by lecturing our children on how they should live. There will, however, be teachable moments that allow us to share understanding with children about the principles. The timing of this will be dictated by circumstances, and there will be many opportunities to learn. When Cory came into the kitchen in a rage, Kevin didn't react or try to change him. He had faith in Cory's innate ability to come out of it. Kevin's recognizing and respecting that Cory was not open at that point cut short what could have been a lengthy negative situation. When we teach health and understanding to our children, we give them the greatest gift we can—the gift of security. A secure child is a happy child, a child who learns easily, a loving child.

Practice what you preach. The best parenting is living in your own mental health and allowing yourself to have human failings. Accept yourself and your children for what makes all of us human.

20

LOW MOODS CAN DISTORT OUR PERCEPTION OF CIRCUMSTANCES

On one particular Tuesday, I felt like nothing was going right for me. I was supposed to have flown to Michigan on Saturday to give a seminar, but had been forced to postpone my trip because I'd been sick with a virus. My illness stubbornly persisted for four days, along with a fever of 102 degrees. When Tuesday came, however, I finally felt good enough to travel, but it seemed like I was running into one obstacle after another.

It was a comedy of errors. When I caught my flight, I hadn't eaten lunch, but the airline didn't serve anything that looked remotely edible. I arrived at the Detroit airport expecting to rent a car to drive to Lansing, but for some reason there wasn't a single rental car left in the entire city of Detroit. After a series of fruitless phone calls to every car rental agency in the city, my hunger was growing and my energy level was dropping fast, along with my patience.

I checked with travelers' aid to see if they had any suggestions, and they recommended I ask the shuttle service. The woman who worked for the shuttle service told me it would cost me $275 one way for a 100-mile ride. *What a rip off,* I thought angrily. *Maybe I should have stayed home in bed.* I was really starting to hate the Detroit airport.

I frantically tried to get hold of the secretary where I was to give the seminar to see if she had any ideas. After twenty minutes of trying to track her down, I finally reached her. She didn't know what to do either and said she would look into it. "Call me back in twenty minutes and I'll come up with something," she said. "In the meantime, see if you can get a flight to Lansing."

"Okay," I said wearily and hung up the receiver.

Suddenly, I realized that my bag was nowhere to be found on the baggage claim carousel, and I started to panic. *Is everything going to go wrong today?* I wondered, as I anxiously scanned the carousel for my bag.

Finally, I realized that I had been standing at the wrong baggage claim carousel and discovered that my bag had arrived fifteen minutes ago at another carousel. At this point, I felt absolutely brain-dead. *I knew I should have stayed home,* I thought wearily. *I just can't think clearly.*

The airport took on a seedy look. The floors looked dirty; the trash bins were all overflowing. Everyone looked unfriendly, and all I could think about was being in my own bed.

I schlepped my bag up the escalator to the ticket counter, only to find the line 150 people long. I decided to call the airline on the phone because I knew I'd never make it through that line. Sure enough, the reservation agent told me they had a flight leaving in twenty minutes and there was one ticket left. *Yes! My luck is changing,* I thought triumphantly and momentarily allowed myself a ray of hope.

"However, sir," the reservation agent continued, "I'm sorry but you'll have to buy the ticket at the counter because it is so close to flight time." My heart sank as I looked at the long line in front of me.

I didn't seem to have any choice, so I got in line as it moved along at a tortoise's pace. I could see that there was no way I would reach the counter in time to buy the ticket, run to the gate, and make the flight. At the same time, my stomach was pleading with me to have something to eat—it was now 5:00 P.M.

Out of the corner of my eye, I spotted a Northwest employee walking by the line. "Excuse me sir," I said. "Is there any way I can move up to the front of the line? If I don't, I'll miss my plane."

"No, but you can buy your ticket at the gate if it's a one-way ticket," he said confidently.

"Great! Maybe I can make the plane." I ran to the gate, which was ten minutes from the ticket counter, noticing all the food kiosks on the way as I ran pulling my heavy bag. I arrived at the gate out of breath, truly exhausted, and starved.

"Can I still catch the 5:20 to Lansing?" I asked the gate agent.

"That flight is 50 minutes late," the gate agent replied, "but you can't buy a ticket here. You'll have to go back to the ticket counter."

"But I just came from there and they told me I could buy my ticket here," I said, confused and disheartened.

"I'm sorry sir, but we can't sell tickets here no matter what they told you. You'll have to go back, but if you run, you can still make it since the flight is late."

The Fork in the Road

Still schlepping my bag, I ran up the broken escalator, past the food kiosks, and up to the ticket counter with the long line. I finally made it to the counter and bought my ticket with enough time to just get to the gate and grab a quick sandwich. The thoughts that passed through my head at this point are mostly unprintable. They were so hostile that I couldn't help but notice how low my mood was.

I momentarily considered that my illness, hunger, and state of mind might just be contributing to how I was perceiving all of this. I quickly dismissed this thought, however. My feelings of anger and frustration with the situation were certainly justified, I insisted. But my doubts about my thinking and perception began to nag at me. After I finished my bagel sandwich and my hunger began to subside, those doubts about my perceptions came back, this time with more credibility. *Maybe my mood is off,* I considered.

By the time I was on the plane to Lansing, I had gained enough emotional distance to almost be amused at my plight in the Detroit airport. I pictured myself telling the seminar group about it the next day as a great example of how low-mood thinking affects decision making. My mood lifted as I got closer to Lansing and disappeared when my contact picked me up at the airport. It now all seemed like a bad dream.

The Principle in Practice

Whatever the cause of our low moods, they usually temporarily skew our perception of our circumstances. When we're in a low mood, we momentarily forget where our experience is coming from—our thinking. Instead, it appears that we are totally a victim of circumstances, which is how I felt at the Detroit airport.

When this happens, our negative emotional reactions set off a chain reaction of bad decisions, an inability to see solutions, and a negative distortion of circumstances.

When I returned to the airport a few days later, rested, healthy, and in a good mood, everybody seemed to be very helpful and even the surroundings looked more appealing. It was amazing to me that even though I know what I know about how thinking and low moods can distort our perceptions, I can still get caught up in the illusion of my momentary thinking and get totally fooled by my perceptions.

Think of your moods as a kind of "internal weather." Just as the weather changes on a daily, even hourly basis, so too our thinking fluctuates in its quality. Our thinking shifts moment to moment for a variety of reasons—lack of sleep, hunger, illness, disappointment, random shifts in thinking. The cause of this fluctuation in our thinking is unimportant. The important thing is that we remember where the source of our experience is—our thinking. When we do, we will realize that our low moods are creating perceptions that are no more real than a mirage of an oasis in the desert. If we know it is a mirage rather than a fixed reality, we won't spend a long time chasing after it. See your moods as a mirage, and you'll protect yourself from a lot of self-inflicted harm.

See your moods like internal weather affecting your moment-to-moment experiences. Learn to wait out the storms, before you make important decisions.

21

TRUST IN THE POWER OF A CALM MIND

Have you ever been so panicked about losing something or remembering something that you couldn't get the thought to come to mind? When we become upset and fearful, it almost seems as if our IQ goes down, and the most simple and obvious solutions or memories become blocked.

I remember once when I was at a dance in high school, I was about to introduce a friend of mine to a girl I was really attracted to. I was so nervous and insecure in her presence that I couldn't even remember my friend's name! My face turned bright red, and I never wanted to see that girl again. *Oh God,* I thought, *she must think I'm really stupid!* I had known my friend for three years and saw him nearly every day, yet under pressure and in a state of insecurity I was unable to remember the most basic piece of information about him—his name. Of course, as soon as the girl left the room, my friend's name came back to me.

If I could forget my friend's name, you can imagine how easy it would be for you to forget something important or not see a solution to a problem when you were feeling under pressure. A client of mine named Craig recently recounted a rather painful incident:

"I found just the right diamond ring for Kris," he said. "It cost me an arm and a leg, but I really wanted it to be a special surprise. I decided to take it with me on our vacation to Florida and pop the question when we got there at just the right romantic moment. But then, the moment came one night and I panicked. *Where did I leave the ring?* I thought. *What if I can't find it? She'll be really pissed if she knows how much I spent and lost it. Oh no! I've blown the whole thing.* Craig searched high and low, through his luggage, the condo, and his truck, but to no avail.

Kris sensed Craig's panic and suspected that something serious was wrong. *Could he be thinking about breaking up with me?* she thought. *Maybe he wishes we hadn't come on vacation together.* Soon they spiraled into an argument fueled by Kris's suspicion and Craig's obsession about the ring.

The Fork in the Road

Craig kept thinking, *If she only knew why I was upset she would understand, but I wanted it to be a surprise.* Finally, at 4:00 A.M., sleepless in Sarasota, he broke down and told her the truth. She was so touched and emotional that she told Craig, "The ring's not important, Craig! The only thing that matters is that you want to marry me!"

At that moment, Craig remembered what he had learned about thought—that *when you calm down and let go of a problem, the solution automatically comes to you through your wisdom.* He and Kris cried, and Craig drifted off to sleep, assured that somehow it would all work out.

Then, at 4:30 A.M., Craig sat upright in bed and a thought struck him like a lightening bolt—*The ring! That's where it is!!! I put it under the console of my truck so no one would find it! I can't believe I couldn't remember the most important thing in my life!* He flew down to the truck, ripped out the console, and sure enough, there was the ring, right where he'd put it. He was so full of relief and joy that he ran upstairs, got down on his knees, and formally proposed.

The Principle in Practice

The moment we recognize our unhealthy thinking—in Craig's case, panic—we find ourselves at the fork in the road. Now is the time to either trust in the power of our wisdom or head back into the high-pressure zone of worry, fear, panic, and obsession. Through Kris's understanding and Craig's recognition that his panic was blocking his mind, he was able to let go of his dilemma and fall asleep. Then, in his sleep, the simple and obvious solution floated effortlessly to the surface.

Think of your mind as a pool of water with silt on the bottom. If Craig threw the ring into the pool and panicked, his panic would stir up the silt. The ring would still be in the pool, but obscured by the murky water. All the flailing around in the pool wouldn't turn up the ring. But letting go, doing nothing, and trusting that the answer was already there and would unfold would allow time for the silt to settle, exposing the ring in the clear water. So it is with your mind.

When you trust in the power of your own wisdom, you can let go of unhealthy thoughts like worry and panic. Without effort, the solution or memory will be there, where it has been all along—inside.

A quiet mind is like a clear pool of water.

22

SETTING NEW SPEED LIMITS

Karen was always a kind, compassionate, and patient person. She hated conflict and found it difficult to deal with people who were angry. As a child, she was fragile and sensitive to others. As a wife and mother, she tended to be the peacemaker in her family and a buffer between her two children and her husband, Tom.

Tom was on the other end of the continuum from Karen. He was very aggressive and short-tempered and tended to blame others for his difficulties. As a police officer, he interacted aggressively with people all day long. When he came home at night, he tended to operate in the same aggressive style with Karen and the kids, though he knew he was often too harsh and impatient.

Karen spent the first twelve years of their marriage being patient with Tom, mediating between him and the kids, and essentially living in fear. As a defense against all that fear, she developed a busy mind, constantly thinking about how to resolve differences between Tom and the kids and how to keep him happy so he wouldn't blow up. One day she realized she had lost her love for Tom.

When she allowed herself to acknowledge her indifference, she became even more frightened by the implications of this awareness. Should she get a divorce? How would this affect the kids? What would her family think? Finally, one day she marshaled her courage and said to Tom, "I don't want to do this anymore."

"Do what?" Tom responded, confused and shocked by her tone.

"I can't live with your anger, your blame, and the constant tension in this house. I don't know what I'm going to do, but I just know our relationship has to change." Karen was surprised at how calm and confident she was in saying these words. "I think we should go to counseling," she added.

"I think that would be good for you, Karen," Tom said, obviously avoiding her point that they should both go.

"That's fine," she said. "I'll go without you." Karen felt resolved about changing her life. It really didn't matter if Tom came or not. She just didn't care anymore.

When Karen first met with her counselor, she was upset, confused, and frightened. She didn't know whether to leave Tom or not. Her counselor recommended she not make such a decision until she had her own bearings back.

Over the next few weeks, Karen regained her calm and knew somehow that things had permanently changed within her. She would no longer be the object of abuse. Interestingly, at the same time Tom quit blaming her, and even though he would start to yell at her or the kids from time to time, he would stop himself.

This was much more pleasant for Karen and the kids, but she still didn't feel love for Tom. She moved from feeling indifferent and fed up to nonjudgmental. She began to see Tom as this new person whom she was getting to know for the first time.

The Fork in the Road

One night Karen had to work late. She arrived home later than she had expected and was greeted at the door by Tom's rage.

"Where have you been?" he demanded. "Your children have been completely out of control. Tina's throwing a temper tantrum and Jeremy left the house and I don't know where he is. If you'd been here, this wouldn't have happened!"

Karen was amazed that she didn't become frightened or intimidated by his anger. Her new understanding was her protection. She remained calm; she didn't feel compelled to busy her mind with solutions to "fix" the situation or Tom, as she would have in the past. It was clear to her that he had relapsed into his old behavior and had simply lost his bearings. She said, "Tom, I don't think this is the time to talk. You're obviously upset and I won't talk to you when you're blaming me and others for your anger. Let's talk about it when you've calmed down." She was firmly grounded in her resolve.

Tom responded furiously, "All right then, I'll just leave! That's what you want anyway." He ran upstairs, packed a bag, and slammed the door as he left.

Again, Karen was amazed and delighted that she didn't get frightened. She felt calm and centered. She knew everything would work out eventually if she just *stayed in this powerful place of calm.*

Her daughter, Tina, was frightened by all of the tension and was crying on the sofa. Karen sat next to her and comforted her. "It's all right Tina," she soothed. "Dad is just upset for the moment. He'll see things differently when he calms down. Let's rent a video and have a nice time, just the two of us."

"Okay, Mom," Tina said, surprised at her mother's calm response.

Shortly after, Jeremy came home from his friend's house and announced he was going to stay with his friend that night.

Karen said, "I think that might be a good idea, Jeremy. You and Dad need a time-out from each other, so you can get your perspective back."

Karen and Tina spent a very special night together. About 10:00 P.M., Tom arrived back home. He walked up to the bedroom without saying anything. Karen knew it would be best to wait till the morning to talk to him.

The next day Tom woke up in a very remorseful mood and said, "Karen, I'm really sorry about last night. I don't know what got into me. I think I just don't know what to do with all the changes, and I just jumped back into my old behavior. Let's go away for the weekend, just the two of us. Would you like that?"

"I'm not sure, Tom," Karen replied. "Let me think about it." She could hardly believe she said this. In the past, she would have jumped at the chance to make things right with Tom. *From now on,* she thought, *I want to take all the time I need to make decisions based on my wisdom, not my old habits.*

After a while, she noticed that she actually thought going away sounded like a good idea. She began to look forward to the weekend. After a couple of hours she said to Tom, "Yes, I do want to go with you. Where do you want to go?"

"How about that bed and breakfast we went to for our first anniversary?" he suggested.

"Sounds great to me."

"Great. I'll make all the plans," Tom said, delighted at her response.

That weekend, much to her surprise, Karen fell back in love with Tom. Just when it really didn't matter to her how it would all turn out, it all seemed to fall into place. Tom agreed to see a counselor to get help with his temper, and they got a fresh start in their marriage.

The Principle in Practice

It seems to be a common occurrence in relationships that opposites attract. Karen, a sweet and patient person, married Tom, an aggressive and impatient person. Karen needed to learn to be strong and to set limits. Tom needed to learn to be sensitive and respectful of others. Their marital crisis turned into an opportunity for them both to grow.

All human beings develop coping mechanisms to deal with the circumstances that life presents to them. Eventually these coping mechanisms become obsolete and need to change. For Karen, losing her love for Tom was a wake-up call that something needed to change. Her mind had gotten so busy trying to mediate, to fix—all to avoid her fear of conflict and anger—that she had lost her feelings of love. Tom, on the other hand, had used anger to avoid intimacy and vulnerability. Karen's actually calling Tom on his behavior was the best thing that ever happened to him.

The first step for Karen was to *regain her bearings,* to calm her busy mind. From this state of calm she discovered a new sense of strength and resolve. Her fear disappeared and was replaced by an ability to set limits, but without aggression. When she was no longer a hostage to her fear, she could take the risk with Tom to set a limit on his anger.

When our mind is calm, we access our wisdom, which is full of new thoughts, insights, and ideas about new ways of being. To evolve in a relationship, we must access this inner strength, or we will replace one bad habit with another. Let go and discover a new way of being. You might just find yourself falling in love all over again.

Trust in the power of a calm mind. It is truly powerful and gives us the perfect response for any moment.

23

SAFE IN A STRAIGHT WORLD

When Jan was quitting her job as a counselor at a clinic, she needed to refer her client, Jack, to another therapist. Jack understood; however, he told her, "I don't mind seeing another therapist, but I don't want to see a gay therapist."

Jack's words really threw Jan off balance. They had definitely had an excellent rapport throughout their past ten sessions, but unbeknownst to him, she was a lesbian. At first, her mind whirled. *What would he think of me if he knew?* she thought anxiously. *How can people be so prejudiced? Certainly he would know by looking at me that I was gay. Doesn't everyone know?* Her body tensed up as she recoiled from the all-too-familiar feelings of being judged for her sexual orientation.

The Fork in the Road

As her thoughts calmed down, Jan regained her bearings. She had been learning about mental well-being through a series of training sessions offered by her employer. At first, she really resisted the new ideas, because they conflicted so much with her previous theories of psychology. But now it was beginning to kick in. Instead of continuing to obsess about herself and what Jack might think of her, she began to feel compassion for Jack. She realized that his thoughts were the source of his judgment and that didn't make him a bad person. In the past, she would have become insecure and tried to hurry the session along so she could avoid the discomfort. But this time, she calmed down and reconnected with the compassion she had felt for him in all their previous sessions. Later she told me, "I had finally learned to not take my client's thinking personally, no matter how personal it was."

At that point, she looked at Jack and said, "You've already seen a gay therapist."

"What do you mean?" Jack asked, confused.

"You have already seen a gay therapist, because I'm gay," she said softly.

"But you have a wedding ring, and you have kids."

"Yes, my partner and I are married and we have two children."

Jack was really confused now, and Jan felt empathy for him in his confusion. She told him that she could refer him to several non-gay therapists at the clinic, but there were also several excellent counselors who happened to be gay. She also reassured him that she would respect his wishes.

When Jan told me about this incident, she said, "For the first time, I felt safe being confronted with the gay prejudice. I am also a diversity trainer, and there are times when I would feel very uncomfortable acknowledging I was gay. I would worry what people thought, if they would still respect me as a trainer, and so on. But now that I recognize that *it is my thinking about their thinking that gets me into a reaction,* I can change that.

"One of the truly amazing things to me is how I now react to men. I used to feel afraid of even being in the room with a man. I had avoided being with men and had almost no relationships with them. But now I have a sense of love for many men in my life. It's not about 'women-safe space,' as we talk about in the lesbian culture. I don't have to avoid the male half of the human race. Even if someone is prejudiced against gays, I now see that I have a built-in immunity to that type of prejudice. I see that it is their thinking and not who they may be that is the problem, and I don't have to internalize it or take it personally. I know now I can feel safe anywhere."

The Principle in Practice

We live in a world full of prejudice. To the extent that people believe in their preformed ideas of other people based on sexual orientation, race, age, religion, nationality, political persuasion, and physical characteristics, they are prejudiced. That means all of us have prejudiced thoughts. But if we can recognize those beliefs as thoughts, we can be free of them and not let them get in the way of knowing other human beings.

Jan realized that *her reaction* to other people's prejudice about her sexual orientation was *her* problem, not theirs. She realized that other people's thoughts had nothing to do with her, unless she took them personally. By recognizing her thinking, she became able to be free and safe in a world of prejudice. The more she has learned to not react to other people's beliefs, the bigger her world has become. And ironically, the more comfortable she is with herself, the more others accept her. Even Jack later realized how narrow-minded he was, probably because Jan didn't react defensively, but with compassion. It caused Jack to rethink his own beliefs.

Don't take other people's judgments personally, and you will feel safe and secure anywhere in this world.

24

DEATH WITHOUT LOSS

How could a very close family suddenly lose one of its members and not feel tragedy, grief, or an enormous sense of loss? At the very least, one might suspect that some form of denial was at play. The following story is about one family who experienced the passing away of their husband and father at such a deeply spiritual level that grief was not a part of their experience.

On June 21, 1998, the family was on one of their many weekend camping trips in a state park in Minnesota. At about 3:00 P.M., Will, an avid runner, decided to go out for a run.

It was 4:15 when Charlotte started to prepare supper at the campsite. Nervously, she wondered why Will hadn't yet returned from his run, but she figured it was probably that knee that had been bothering him lately. At 4:30, however, she began to get really concerned, and at 5:00 she sent her children Eric, age eight, and Mary, age fourteen, on their bikes toward the trail where Will had gone for his run.

When Eric and Mary reached the trail, it was roped off with yellow plastic police tape, and several park rangers were blocking their path. "Sorry kids," one of the rangers said. "You can't go there. Someone's been hurt and an ambulance is on the way."

Charlotte arrived a few minutes later and said to the ranger, "I think that may be my husband. Can you let me through?"

He motioned her through, and she saw a body lying on the ground covered with a blanket. Instantly, she knew it was Will by the jogging shoes that were sticking out of the blanket. She hoped he had just been hurt, but she could tell from the look on the ranger's face that he was dead. She forced herself to form words.

"Is he dead?"

"I'm afraid so, ma'am," the officer said sympathetically.

Immediately, Charlotte felt the deepest pain and shock she had ever experienced. Her heart felt like a knife had run through it, and she began to scream and cry uncontrollably. They wouldn't let her see Will's face till the coroner arrived and examined the body.

At one point, she realized her kids were waiting at the end of the trail. Charlotte knew she had to regain her balance to be able to tell them, but she couldn't imagine where the strength would come from.

Somehow, she managed to tell her children that their dad was dead, and the ranger took them to a cabin where they could have some privacy. Charlotte knew she had to stay calm for Eric and Mary and not go back into that feeling of excruciating pain. She had to go back to the campsite, pack up everything, and start the two-hour drive back home. While at the campsite, Charlotte recalled the thought, *what a huge transition for all of us.* Charlotte and her children were overcome with sadness as they started their journey home. All they wanted to do was get there. Their mouths were dry and their stomachs were upset.

On the edge of collapsing, Charlotte began to talk out loud to

Will as she drove. "Will, we need everything we have to get there. Please help us get through this," she pleaded.

A feeling of gratefulness washed over Charlotte when she noticed the sun. It had been there all along, but it was the feeling of gratitude that allowed her to see it. "I'm so thankful that the sun is out," she thought, remembering how stormy it had been most of the weekend. "I'm glad we don't have to make this drive in dreary weather." She felt as if she had grabbed onto something deeper within her that she had never felt before. And there was the strength she needed inside her that she didn't know she had.

About 9:00 P.M., Charlotte arrived home and began calling her family, but the only one she could reach was her sister, who was in Montana, fifteen hundred miles away. Then she called my wife, Michael, and me. When she got our answering machine, Charlotte feared being alone for the night.

Fifteen minutes later we returned home, heard the message, and called Charlotte immediately. Later she told us, "I realized how perfect it was that my family was out of town and you both were able to come over that night. They would have been emotional and distraught. It was just what I needed, people who could come with love and compassion without the heaviness of grief. You acknowledged our emotional state without pity or commiseration."

The Fork in the Road

Before our arrival at her house, Charlotte got a call from the Red Cross asking her if she would be willing to donate Will's eyes and tissue for transplants. It was then she affirmed to herself that

Will was no longer in his body; his spirit or consciousness was somewhere else. Surprisingly, it felt right to say yes to the Red Cross worker, though she knew that in life, Will would have probably freaked out at the thought.

"Somehow," Charlotte told me, "I just felt in my heart that it was okay with him. I felt like Will was communicating with me through my heart, my deepest feelings. Then, the next morning, it was very quiet as I sat in the backyard. I felt Will's presence so strongly as I struggled with whether or not to cremate his body. And again, I clearly felt him giving me permission to do whatever would be the best way for me and the kids. It was as though my relationship with Will had moved to a totally new level with his death. Many of our little habits that had bothered each of us seemed to disappear, and all that was left was this deeply spiritual love for each other. My love for Will actually grew with his death. I was grateful for the transformation that was occurring in our relationship.

"I actually feel closer now to Will than I ever did in life, even though we were very close while he was alive. His death really hasn't been a loss for me. It has been a gain. That is really been the crux of the matter, my realization that there is no loss. Sharing this with others has been extremely helpful to them."

This is especially true for Eric and Mary. Will is still a strong presence in both of their lives. They talk to him often, and they write letters to him about their feelings. At Will's funeral, Mary read a letter she wrote to her dad, the most beautiful and wise letter I've ever heard. It helped everyone at the funeral to heal their grief. The children have gone on with their lives and are joyful, though at times they feel the loss and sadness.

Charlotte acknowledged to her children that Will was still part of the family from the very outset, as they made their way home from the park where Will died. Actively communicating with him and maintaining their love for him has healed Charlotte and her children. They are honest with themselves and with one another, and they talk straight from the heart. For all of us who know them, the whole family has been a model for what is possible even with as sudden and as great a loss as this. It has personally given me a great sense of relief and comfort to see that it is possible to survive a great loss and actually grow in the process. Through Will's death, Charlotte and her children have taken a huge leap in their spiritual and psychological growth.

The Principle in Practice

How could a very close family possibly deal so positively with a death that was this sudden? The answer is that Charlotte and her family had a very solid understanding of the principles of mental well-being. Through this understanding, they were already a living example of a healthy family. They lived a good life together and shared many deep feelings of love and joy. Their story shows a deepening of their understanding of the principles of Thought, Consciousness, and Mind.

Charlotte's understanding of the principles deepened with Will's death. She had experienced both sides, living in the moment and letting her mind fill with thoughts of fear and loss.

Her response was not one of denial, but of insight. She chose to live in the moment.

Charlotte's deep insight enabled her to recognize the difference between being connected to the deeper experience of Mind and being caught up in a fear-based circular thinking process. The difference became profoundly obvious to Charlotte. Living in the moment was healing; she felt calm, safe, grounded, and able to appropriately respond to her needs and the needs of her children during this emotional time. When we are in the moment, we experience our connection to Mind. Mind is the source of all life. It is the energy we are all part of and connected to. When Charlotte lived in the moment she could experience Will because he too is part of Mind.

At this pivotal time when the other side of the coin was intense pain, regret, and loss, Charlotte knew she did not have the "luxury" of indulging in previous negative thought habits. The consequences were too painful. She did not want to experience the negative thinking or take her kids there.

Many times we know that certain negative thinking habits are not beneficial to us, but we indulge anyway, inviting those thoughts to take residence in our mind. Charlotte had taken a jump in understanding; she stayed steadfast in her efforts to live in the moment.

Not all of us will be able to handle such a life-changing experience as gracefully as Charlotte and her family; but through their story we are given a vision of what is possible. Through this hope and vision we can trust the power of living in the moment; we can understand that letting our thoughts go and not dwelling on any of them is the cleansing agent that will transform our grief and bring us back to the moment. Then and only then will we feel our connection to Mind, the source of all life, and experience a deep connection to our loved ones, even in death.

Mind, the life-force, is eternal; it can neither be created nor destroyed. What we do with this life-force is our own personal creation. At a very deep level, Charlotte and her children see that in actuality, Will has not died. Only his physical form is gone. His spirit and his consciousness live on and remain a part of their lives. That is why they can experience Will's death without loss.

There is no reason to fear death. It is a powerful transformation, but it is not an ending.

25

"DEAR DADDY, DON'T WORRY, WE'RE FINE"

Mary was fourteen when her dad died suddenly of a heart attack. At first, Mary was in a state of shock and disbelief; she was numb and tired. Then her mind started to speed up with memories and the implications of his passing. She was afraid of the pain that might follow the shock and dreaded it. *Now what?* she thought. *What will I do without my dad? We were so close. I won't be able to go to work with him. I won't be able to ask for help with my homework. And all those special times we had going to the coffee shop together. . . .* As Mary began to grasp the reality of her dad's death, her thoughts spun out of control.

The next couple of days she continued in this state of numbness, shock, and spinning thoughts as she tried to make sense of this huge change in her life. She later told me, "I started to think of what Daddy might be going through. Sure, it was a shock for us, but what about him? Dying *really* has to be a shock. I prayed for him and asked God to help him through this confusing time."

The Fork in the Road

About three days after her father's death, Mary felt a shift inside of her. "All of a sudden," she said, "I could feel his presence all around me. I could feel Daddy's feeling in my heart. I know this might sound weird to a lot of people, but this is what I experienced. The way I felt when I was with my dad when he was alive was what I could feel that day. I was given some information, a knowing that everything would be all right. I began to feel really happy."

I was around Mary during this time and witnessed her shift. She was joyful, even though she had her moments of sadness and tears.

"I just wanted to celebrate my dad's life," she said. "I could see that death happens so we can wake up to how we are connected to God. Through my dad's death I have come to understand this and have developed a close relationship with God. This has been good for all of us—Mom, my brother, Eric, and my whole family. It has shown me and others just how precious life is and all the love we have for each other. I truly believe this, and I really trust my instincts. This has been an incredible gift for all of us."

She recently said to me, "I would have liked to have him continue as my dad, but that will get taken care of in a different way. I feel his presence often. If I am sad or going through some difficulty, all of a sudden he is there, and I feel calm. His presence comforts me even in his death. I don't feel he is gone. He's just gone physically, but his spirit is still here. If I could have ten more minutes with him, I wouldn't really need to, because I feel that the relationship we had was complete. We said everything we wanted to and really loved each other."

I asked Mary what advice she would give to someone who had experienced the death of a loved one. She shared the following ideas with me:

1. Pray for those who have died. They are going through a big shock too. They need your thoughts and prayers to get through the transition.

2. Keep them in your life. Think about them often, pray for them, talk to them, ask for their help. Keep them alive in your mind. Keep their pictures around, write to them, don't forget them.

3. Don't be so sad. Sadness comes from not really understanding death. It is just a change from the physical to a spiritual existence. It is a returning to a oneness with God.

4. Don't regret anything. It won't do you any good, and they wouldn't like you to regret anything either.

Mary truly did celebrate her dad's death. At the funeral, she courageously got up in front of the church and gave a eulogy. She had never given a speech before, but she moved everyone to tears and uplifted their spirits. Here is a portion of the letter to her dad that she read at his funeral:

Dear Daddy,

Don't worry, we're fine. And I'll never ever ever ever ever, for one single second of my life, forget how special you are to me. I love you, Dad. I hope you can see me and Eric grow up from where you are in heaven. Don't be sorry about anything. I know this is such a surprise for you, and it must all be so different. But this is the way it is. . . . I probably have no idea what it's like up where you are, but I can certainly feel that you're happy and comfortable. That's so good for me. Really good. . . . You know what? I'm happy that the fourteen years I spent with you were such great ones. Okay? I want to let you know that. And please don't be sad, Dad. I know you're still alive—not your body, but you. Your spirit. Isn't that great that dying isn't the end? I'm so glad. Dad . . . I know you're in shock. It was so sudden and shocking. But you will be okay. Trust me. It'll sink in and you will get used to it. So will we. . . . Well, I send all my love to you and you are still and will always be my dad. I'll never go without you, and I know you're smart enough to know that. So don't worry. I'll take care of myself and so will Mommy and Eric. Here, I have a quote . . .

In the darkest hour, the soul is replenished and given the strength to continue and endure.

And I believe that so much. I believe something has touched me and given me the strength I have. I never thought I would take this so well. I always imagined I'd be bawling for days. And maybe right now, inside I am . . . but I'm learning so much from all this that I'm taking this all in the best way I could. I'm doing good.

Love,
Mary

The Principle in Practice

While Will was still alive, Mary and her family lived in a deep feeling of love. This love continues. When we understand that death is only the death of the physical, we can continue our relationship with the person in our lives who has passed on. Consciousness, spirit, soul, energy—whatever you want to call it—is eternal. It is only in our thinking that we can get momentarily trapped in the illusion of a loss.

This is not to say that sadness is wrong in any way. Sadness, confusion, anger, and regrets are all thoughts and feelings that will pass through our minds in a time of loss. However, if we don't see beyond those thoughts to a greater truth, we will become stuck in a painful grieving process. When we experience these thoughts and feelings with understanding, they will gradually lessen, like laundering a dirty shirt in clean water. Mary and her family were able to lessen their pain by living in the moment and having faith in the power of this process. If we, like Mary, do reach the deeper understanding of life, we will celebrate our loss.

Love is eternal and cannot be destroyed through death.

26

TEACHING PEOPLE TO FISH

Helen has been helping people ever since she was a teenager. When she was seventeen, she joined Vista Volunteers because she wanted to make a difference in the world. In that program, she helped the poor meet their basic needs of food and shelter, and she assisted them in getting legal help. Later on she worked with women offenders who were being released from prison. Because of her energy, commitment, and enthusiasm, she moved up in the system. Ultimately, she got her degree in social work and moved to California, where she worked in a county human services division.

After a few years of this, however, Helen began to tire of the inherent limitations of her work; she saw that the help she gave to her clients was only temporary. She realized that her clients didn't have the capability to help themselves and were dependent on a system that was more their adversary than their ally. But she remained very driven and seemed to define herself through helping women whom she saw as helpless victims of the system and society.

Not surprisingly, Helen began to show signs of burnout from all the stress she was under. Her health began to suffer, and her husband left her. She began to fantasize about leaving social work and becoming a therapist. Maybe if she helped people to change inside, they would be able to help themselves in the outside world. They would stop being dependent on an inadequate system.

In search of this new way of helping others, Helen went to graduate school in psychology, but she became disillusioned before she even got her degree. She could see that what they were teaching her was also filled with hopelessness. It was about labeling people, giving them a prescription or a technique, and changing them from the outside in. The process was really much the same as it had been in her social work, but this time with a psychological twist. She learned to see people as damaged goods, many of whom would be hopelessly mentally ill and dependent on therapists for the rest of their lives. *Was there any hope that people could actually become healthy?* she wondered. *Or would they just have to cope with their illness or addiction?*

The Fork in the Road

Toward the end of her graduate program, Helen attended a workshop at the urging of a friend and colleague. On the first day the workshop leader said, "Just listen today. Try not to figure out this new information or fit it into what you have previously learned. Listen for yourself." These words hit her at just the right time. After her grueling graduate program, it felt like such a relief to just sit and listen rather than analyze.

Helen told me about what happened next.

"That afternoon," she said, "they taught us about the three principles of Mind, Thought, and Consciousness. It was as if a light turned on in my head! All of a sudden I could see why the outside-in models of giving people food, shelter, medication, and legal assistance actually stripped them of their power. I saw that the missing link in psychology and social work was the ability to truly help people learn how to do things for themselves—to learn *how* to fish instead of just being given fish. They needed to know how their minds worked to create the life they had. It was so simple, but it made so much sense. I had been seeing everything totally backwards. It was the biggest 'ah-hah' I had ever had.

"The next day I heard about how each person is born with resiliency. Was that true for me too, I wondered? Maybe that is why with all I have been through, I always land on my feet. It gave me hope to see this concept and know that I could help people by leading them back to their own wisdom."

The next day in her graduate program's therapy group, Helen began to see her clients' health instead of all their symptoms and labels. It was as if she were looking at them with new eyes. If she just showed them what she had begun to learn about innate mental health and how thinking works, would they feel as good as she did?

At the time, Helen was working with some of the most difficult people in the system—the chronically mentally ill, prostitutes, substance abusers, prisoners at the county jail. But by teaching them the principles of mental health, she began to see her clients quickly and easily make transformations that she and her supervisors had never dreamed possible.

For example, she worked with a prisoner named Juan, a chronic amphetamine addict who robbed to support his habit. He had been through substance abuse treatment eight times with no success. Now he had discovered an inner calm that had quenched his desire for drugs. He once said, "My thinking got me into prison, and it is my thinking that will keep me out of prison."

And soon her colleagues were seeing dramatic changes in Helen as well. They would say to her, "Helen, what happened to you? You look so much younger and happier. You used to be so burned out. What happened?" She was almost reluctant to share with them what she had learned, because it sounded so simple. But since then, hundreds of professionals in the Santa Clara County social services system have been trained in this understanding, and Helen is now doing training most of the time. She has taught these principles in jails, schools, mental-health facilities, and chemical-dependency treatment centers. Finally, she has learned to teach her clients how to fish and has fed herself in the process.

She told me recently, "I now know how to really help people. This is what I have always wanted to do, ever since I was that thirteen-year-old girl who wanted to make a difference in the world. I no longer feel driven in my work or stressed. My work has become effortless, now that all I have to do is point people in the direction of their own health. The message is simple, clear, and positive. What more could I ask for as a helping professional? And my own life is wonderful as well—my relationships, my health, my enjoyment and sense of purpose in life."

The Principle in Practice

When helping professionals see the human potential in people, their job becomes easy. This is true for anyone who works with people—from parents, teachers, and police to managers in a business. We social workers, therapists, and health-care providers often feel the burden of responsibility to help others resting on our shoulders. Seeing that people have the power to change themselves doesn't mean that we abdicate our role as helpers, but it does relieve us of a tremendous burden. Like Helen, all we need to do is direct others to their core of health and show them the *how-to* of change by sharing the principles of Mind, Thought, and Consciousness. The rest is up to the person who wants to change.

The reason so many people in the mental-health and addiction fields are burned out is not because of the nature of the work itself; it is because they do not see anything but temporary results at best. Any job would be frustrating if we never got any results. Certainly, being compassionate, understanding, and patient and knowing how to help are still essential elements for anyone who works with people, but what really creates effective and permanent results is sharing with clients the understanding of how to be healthy within themselves and not keeping them dependent on the system.

There is no greater privilege than working in the profession of helping people. Once we understand how to help, we are blessed with seeing people become happy, productive, and glad to be alive. That is truly a gift!

Feeding people fish allows them to eat for a day. Teaching people to fish enables them to eat for a lifetime.

27

BEING IN THE MOMENT: THE SECRET TO INTIMACY

The popular lyric from the 1970s song "Slow Hand" by the Pointer Sisters certainly sounds like it's about sex, but I believe it speaks for women who want their men to slow down and be present in all areas of their relationships. Often in our intimate relationships, men—and women too—are preoccupied rather than present. We are so distracted by the things we have to do or what happened that day that we don't take time to really connect—to be intimate.

What does it mean to be intimate? The feeling of intimacy comes from being present, with a clear mind, with someone you love. It is only when our minds are quiet and calm that our deeper feelings of love, caring, compassion, appreciation, and gratitude bubble to the surface. When two people experience this together, their hearts seem to join as their warm feelings are felt. This is the essence of true intimacy.

Earlier in my marriage, my wife would complain that I wasn't *really* present much of the time. We would hug, but I seemed to have an unconscious timer inside my head that rang "Time's up." When that timer went off, I would push away, and she would feel rushed. The message she got from me was, *I have more important things to do than to hug you.* I wasn't aware that I was doing this; after all, from my perspective, at least I was hugging her!

In hindsight, I now see that the problem with me was that I didn't know how to be in the moment, no matter what I was doing at the time. My lack of focus when I was with my wife or any other person was just another example of my inability to be present. At the time, I had no idea what I was missing. My idea of going through life was to keep a "mental check-off list" in my head: take out the garbage √; do the dishes √; fill the car with gas √; hug my wife √. Once one item was done, I was off to the next one, never really being present or enjoying the moment.

This was a habit of thinking that seemed totally normal to me. In the Midwestern agricultural family I grew up in, "getting things done" was our number-one priority in life. It was more important than taking time to do something really right, more important than relaxing, more important than visiting or taking time to simply be with someone. Once our work was done, we could relax and be with each other, but it seemed to me as though the list never ended.

I remember going to Guatemala as an exchange student when I was seventeen. I couldn't believe people actually took the time to listen to each other, hang out, and have fun. At first it made me very uncomfortable. *Don't these people have anything to do?* I wondered. *Where do they find the time?*

As time went on, however, I began to enjoy the relaxed pace. Guatemalans are really close to their families, friends, and even their co-workers. It was a sharp contrast to the busy, distant, and rushed interactions I was acquainted with in my culture. When I returned to Minnesota, I even entertained the possibility of becoming a missionary priest, so I could return to the slow-paced and intimate culture of Guatemala. What I loved most about being there were the feelings of intimacy that I had missed in my world of to-do's and rush, rush, rush.

As my life progressed, those times in Guatemala began to fade into a distant memory, and I rejoined the "fast pack." Finish college, finish graduate school, start my career, become a success—the treadmill went faster each year. Then my marriage hit a crisis. I thought we had a great marriage. I was confused and yet I knew intuitively my wife was right—something needed to change. We soon discovered that what seemed to be an innocuous habit of "not listening" was undermining our marriage. It was affecting our communication, our intimacy, our evolution as a couple.

The Fork in the Road

I knew that I loved Michael more than anyone I had ever loved. When she announced just how disconnected from me she felt, it really got my attention. I vowed to learn how to be in the present and to listen.

In order to be present and intimate, I first had to see how it was connected to my thinking. Over time, however, I began to see how my mind would wander, how I would become easily impatient while listening, and how I just couldn't be still. As I noticed my thinking in the moment, my mind began to calm down, and I started to really enjoy the moments of intimacy with my wife.

And I don't mean just sex. So many men look forward to sex, because it is the one time in their lives that their minds are in the present moment. But I began to enjoy the moments when we were just sitting on the couch together or lying in each other's arms. It was so sweet! Being in the moment has become the most precious gift of my life.

The Principle in Practice

Understanding how to be in the moment and to listen was the turning point of my life. As I began to be present, my enjoyment of everything went up, my respect for other people and theirs for me grew, and my level of productivity actually increased; in short, my life changed across the board.

Being in the moment comes from being in a flow of thought that has the capacity to recognize itself. When I am simply aware of the fact that my thinking is creating my experience of reality, I am able to keep coming back to this moment. Just as the ribbed grooves on the side of a highway let us know that we are veering off the road, the hectic, preoccupied, or unpleasant quality of our thinking will let us know that we are veering out of the moment. When we respect being in the moment, we will recognize our thinking and pull ourselves back into the sane lane; our lives will then be smoother traveling and filled with many more intimate moments.

Value living in the moment, and listen, listen, and listen some more!

28

TO THINE OWN SELF BE TRUE: KNOWING WHEN TO SAY NO

One of the main causes of getting caught in the speed trap is feeling overwhelmed and overloaded with obligations—family traditions (birthdays, holidays, social functions), civic duties, and promises to friends and neighbors. When something we *want* to do turns into something we feel we *have* to do, it is usually a sign that we are on overload and ignoring our common sense and our inner feeling about our own personal needs. We get caught in the trap of deciding what's the "right thing to do" versus *what feels right*. We become torn between feelings of guilt and exhaustion.

My mom is one of the most giving, loving, and thoughtful people I have ever known. She has always made each of her children, godchildren, grandchildren, and great-grandchildren feel deeply loved and appreciated. And for most every occasion, she has always given us a card or gift. Although her actions are heartfelt, at the same time she is very conscious of not wanting to hurt anyone's feelings.

Last Christmas, Mom had to make a very difficult decision. She is eighty-eight, and because my dad died at Christmas two years ago, the holiday season has become a difficult time of the year for her. She wanted to continue buying presents for all of us kids, but she just didn't have the energy. Her inner "feeling" was telling her not to do it, but her "head" was telling her she should buy the presents.

"What would people think?" she shared with my sister one day. "Wouldn't they be hurt? Would they still know I love them?"

"Mom, don't be silly," my sister assured her. "Everyone understands, and we all know you love us. No one could ask for a more generous, kind mother."

To my mom's credit, she listened to herself, instead of her guilty feelings of obligation. No one thought she was selfish, and no one felt unloved. We were all happy that she was taking care of herself at this difficult time. She was the only one who felt bad about it even for a little while. Yet in hindsight, she was very glad she listened to her heart instead of her head. She had a much better holiday, which was the best present any of us could have had.

Many of us try so hard to be a good parent, grandparent, employee, or friend that we often forget to be good to ourselves; we treat ourselves as some sort of second-class citizen. We put ourselves last because we think it is more noble or more loving. The problem is that we "think" it is a better thing to put others first, instead of listening to our wisdom.

Wisdom or common sense speaks to us through a feeling. The feeling lets us know when we have too much on our plate, too many things happening in our lives. This feeling is like the whistle on the teakettle that tells us the water is boiling and it's time to turn down the heat.

When we listen to this inner signal as we go through our lives, we find ourselves in a state of balance. Balance takes us out of the speed trap of life and puts us into a state of well-being. And when we listen to our wisdom, it works out for everyone else as well.

Sara, a good friend of mine, related the following story to me about listening to her own wisdom:

"My daughter-in-law was visiting from out of state for ten days with my grandson. Every night, my other children stopped by with their kids, and everyone would stay for supper. A few nights of this was fun, but then I started dreading their visits. I just wanted to have a quiet, ordinary night, with no more food to prepare."

That Sunday, my friend Sara found herself thinking, *I really hope no one shows up tonight. I need a break from all this activity.*

Just then, the phone rang. It was her son-in-law.

"Hi, Mom," he said. "Are we going to do our Sunday dinner thing?"

"I don't know, John. Let me think about it and I'll call you back."

Sara hung up, tormented by feelings of guilt. *Am I being selfish? Should I listen to my own feelings about this? Am I going to disappoint the grandkids? I don't care, I just can't do it! I can't believe I just thought that.* Her feelings of guilt bubbled up even stronger.

The Fork in the Road

At that moment, Sara had a realization. *I guess that when I trust my feelings, it seems to work out for everyone. And I do tend to do things out of obligation, so maybe I should trust my feelings for a change.* Although Sara had been learning about the principles of mental health for a number of years, the issue of family obligations had continued to plague her.

She picked up the phone, called John back, and said, "I hope you and Kate can understand, but I'm on overload and just want a quiet evening at home."

"Don't worry, Mom. I can fix dinner and clean up," John replied.

"The point is, I just don't want to have all that stimulation, even if I'm not cooking the meal." Sara felt the urge to give in tugging at her, but her inner strength took over. "Tonight's not a good night, John, but thanks for offering. That was kind of you."

With those words, a warm feeling of calm took over. Sara knew she had done the right thing, and she knew that John, Kate, and the kids would be just fine.

The Principle in Practice

As human beings we have a built-in mechanism that guides us in life—*wisdom*. Sometimes what our wisdom tells us seems to conflict with social obligations, rules, and what other people think; however, we know *in our heart* that it's the right thing. The more we learn to trust these inner feelings, the more our lives return to a *state of balance*.

Another interesting point here is that when we all respect each other as well as ourselves and each person listens deeply to his or her own inner wisdom, the group itself is brought into a state of balance. When I trust my wisdom and have the courage to follow it, it may mean temporary disappointment to someone else, but it ultimately turns out to be an important lesson or experience for that person. Besides, I would not be doing myself or anyone else a service by acting only out of obligation and ignoring my own wisdom. So in the big picture, it all works out.

Learning to listen to our inner feelings of wisdom can feel selfish, but actually it is the truest act of selflessness. It takes courage to ignore social conformities and rules and be true to ourselves.

When in doubt, remember the great wisdom in that memorable quote from *Hamlet*, "To thine own self be true."

29

THE QUICKSAND OF PROVING YOURSELF

Mary Jean is a psychotherapist who came to a training session I conducted a couple of years ago. She was very skeptical of this new psychology and fearful of giving up many of the ideas that she had come to believe about herself and helping others. In fact, she was very anxious and suspicious about life in general.

"I was actually afraid of people," she said. "Funny that a therapist would be afraid of people, but I was. I would approach new clients with a feeling of dread and anticipation. Would I be able to help them? Should I refer them to someone who knew more about their problems than I did? What would they think of me? Much of the time, I felt extremely anxious and would have to give myself a pep talk every time I went into a session."

Mary Jean grew up feeling inadequate and less intelligent than her siblings. As a psychotherapist, those thoughts hounded her in her day-to-day work. She would often refer clients to other therapists not because they were appropriate for the kind of problems those clients had, but because of her own feelings of inadequacy. As she put it, "I wasn't really very present with my clients. I spent a lot of time in rooms with them, but I was preoccupied with my own thoughts of how I was doing and how inadequate I felt. I did all of this running around in my head about things, like wondering if I should have another therapist see them instead. I would make mental comparisons between what I said to clients and what I imagined another therapist would say. Occasionally, I would forget about myself and actually help others, but it was rare. Most of the time my mind was very busy."

The Fork in the Road

"In the past, after I was working as a therapist for a few years, I would make many strides in improving my skills and would help many people, but I would also fear that the very next moment I would step into the quicksand and be sunk again. No matter how much I proved I was an adequate therapist and person, I just knew at any moment that I would be exposed as the incompetent person I actually was. For ten years, I was actually aware that my feelings of inadequacy were holding me back, but I just didn't know how to stop them. Learning that the source of my experience is in my moment-to-moment thinking has freed me from those thoughts of inadequacy."

I asked Mary Jean what it was within her that had actually changed.

"I guess I realized that I'm safe," she said. "I don't really have anything out there to fear. It was all in my head; it was only my thinking all along. As long as I remember that, I'm okay. I may still have the thoughts from time to time, but once I recognize them, they vanish. I am able to be calm with my clients and actually put my focus on *them* rather than on *me* and *how I am doing*. I listen to my clients and even listen to my own wisdom and insights, which is where the real power to help my clients comes from.

"Over time, I have come to really believe that I am okay and I am competent, because time and again the thoughts of inadequacy vanish when I recognize them. My true potential has finally come out. Because I see that potential in myself, I have been able to share it with my clients in a very helpful way.

"I don't dread seeing any of my clients now. If I do feel some anxiety, I know it is only my momentary thoughts about my performance as a therapist."

The Principle in Practice

No matter what our profession or occupation, we all have feelings of inadequacy from time to time. Although we all encounter situations where we don't have all the answers or where we *are* actually in over our head, this doesn't necessarily have to lead to feelings of inadequacy. When we are okay with not knowing, we can be confident no matter what we are facing.

Mary Jean had the habit of thinking about how she was doing and judging herself against others. This kept her mind speeded up and unavailable for insights about how to help others. It also kept her mind too busy to really listen to her clients. When she realized where her feelings of inadequacy were really coming from—the thoughts in her own head, and not her lack of expertise or the difficulty of the situation—she was in a position to change.

Like being in quicksand, the harder we try to prove ourselves, the deeper we sink into the feelings of inadequacy. The reason for this is that proving yourself has a fatal flaw built into it—you are not only the person who is compelled to prove yourself, you are also the person who judges your efforts. Therefore, any efforts to prove yourself are doomed to failure, unless you recognize that the problem is your thinking, not who you are.

Instead of struggling to prove yourself, see that you are already perfect, even if you don't have all the answers.

30

TEACHING TO THE HEALTH IN STUDENTS

Angela is an elementary-school teacher. At a time when the dropout rate for teachers in her area is 50 percent, after five years on the job she says, "I am grateful to have a teaching position. I get to see that glimmer in the children's eyes as they are learning and know they have the potential to meet success, and that is so rewarding." This is the story of how Angela came to an understanding of her own mind and how that had an impact on her role as a teacher.

For as long as she could remember, Angela had always wanted to be a teacher. She was a good student herself, always positive and enthusiastic. She was sure her personality would be perfect for getting along with the students and for being a successful teacher. Halfway through her studies, she had her first practical experiences working directly with kids as a student teacher. However, she soon became discouraged when the students didn't respond to her enthusiasm. They acted as if they couldn't have cared less about what she was trying to teach them. After those negative experiences, she considered leaving education. Maybe she just wasn't cut out for the job, she thought.

Angela even left her program for a year and studied in Europe and reflected on what she really wanted to do with her life. When she returned, a close friend who had always been a seeker of truth came to visit. He had studied all kinds of esoteric and traditional philosophies, but he shared with her, "I've found it, the secret to life." He talked about thought and the answer being inside. He tried to explain what he discovered, but it was confusing to Angela. The feeling he had, however, was profoundly deeper than before.

Later Angela went to hear a talk given by the person from whom her friend had learned this new philosophy, a man named Sydney Banks. She was very moved by his talk and afterward walked up to him to ask him a question. As she patiently waited her turn, she had a realization that she later shared with me. She said, "My question was coming from my intellect, and I could see that this was beyond any kind of intellectual understanding—it was in the *feeling.*"

The Fork in the Road

Shortly after that encounter with Sydney Banks, Angela, who by this time had returned to school, started the final practical phase of her education. She approached this stage of her training armed only with the new feeling that she was experiencing and not really knowing why she was experiencing it. All she knew for sure was that for the first time in her life, she felt truly happy, calm, and secure. It seemed as if she was immune to her students' negative attitudes and behavior this time, unlike during her previous student-teaching experience. Whenever her students were negative, her mind would remain clear, instead of getting muddled in trying to figure out what to do with them. The static of self-doubt in her head was gone. As a result, she received the highest possible evaluation for her teaching. The only thing she could surmise from this was that her new *feeling* somehow protected her and engaged the students.

She really enjoyed the next two years of teaching until she got pregnant with her first child. Then she took some time off from teaching to raise her family, two children in all. When she returned to teaching, she had become far more grounded in her understanding of her mind and mental well-being. "It wasn't just that I believed the children were healthy underneath all their problems," she said. "I was unshakable in my conviction. Nothing could faze me. No matter how serious the children's problems were, I was convinced I could reach them, and I did. I had learned to see beyond their behavior and problems, beyond the labels of 'attention deficit disorder,' 'disruptive,' and 'school phobic,' and saw only their health and human potential. Most of my colleagues were extremely proficient at diagnosing problems and developing strategies to deal with them. I was just quietly seeing my students' health, staying in touch with my own health, and getting remarkable results. Children would come into my class with serious histories and leave being themselves—capable, secure, happy, and successful."

Angela noted that by her focus on their health rather than their deficits, the children became calmer. And as they became calmer, they began to do several things:

1. They seemed to be able to access previous learning that they had been unable to access before.
2. They rekindled their desire to learn and saw themselves as capable of learning.
3. They rewrote their personal histories in a more positive light, thus changing their image of who they were. They also wrote a positive image of their future, believing that what they could do and accomplish was far greater than they had previously believed.
4. Gifted children also had many problems, but they had the burden of high expectations to continue to succeed. They also had previously tied their self-image to the strength of their intellects. Once they realized that their well-being was not contingent on how smart they were, they were able to round out their intellectual development with emotional development, athletics, and other interests.

Angela shared with me how seeing the health in her students helped her as well. She said, "I feel privileged to witness the talents and assets of children coming forth, instead of feeling overwhelmed by all their problems. That's why I got into teaching in the first place—to see the emergence of each child's unique development. What a gift!"

The Principle in Practice

When Angela first began teaching, she was armed only with her enthusiasm. Enthusiasm, like positive thinking, pales in comparison to a deep understanding. Understanding the principles that govern life has lasting power. Angela's gift as a teacher comes from the discovery within herself of a *feeling*. She discovered that when her mind became quiet, she filled up with gratitude, love, and compassion. As her capacity to quiet her mind deepened, she became rooted in the core of her being. All she saw in all of her students was truth and beauty.

As a teacher, Angela tapped into her protection from all of these circumstances by realizing that it was her thinking that created her experience of whatever she encountered. Her inner presence and calm are greater assets than all the curriculums and techniques in the world, because if she didn't have that inner calm and knowing, she wouldn't have interested students. Understanding how to put children at ease in the learning environment has been the missing link in education. Interestingly, this understanding begins with the teacher, as it did with Angela.

Teach from love and see the miracle of health in every child.

31

THE POWER OF FORGIVENESS: SEEING LIFE FROM A NEW VANTAGE POINT

For most of his life, Henry had a chip on his shoulder. The oldest of eleven children, he spent his childhood helping his mother deal with his alcoholic and dominating father. As soon as he was old enough, though with a great deal of guilt, he escaped the turmoil of his family situation and moved on to find a better life, as far away from his father as he could. But no matter how far he went, he was always argumentative and judgmental and would often pick a fight. The rage of his childhood was inescapable.

Henry also became a seeker of truth. It was the 1960s, and so he engaged in the seeking of those times—Eastern religions, pop psychology, and eventually alcohol abuse. Along the way he met a lovely woman, Tracy, and they married, but Henry still carried his anger with him and took it out on the one person he loved the most. He and his wife were constantly bickering, mostly at his instigation, and they would often separate. Without realizing it, Henry was becoming his father.

One day Tracy met a very unusual man. This man was very ordinary in some ways—he had been a welder, had a wife and two kids, and had the normal troubles of most people. But one day, in his forties, he unexpectedly had a profound realization. When Tracy met Sydney Banks, he was at the house of one of her friends, speaking to a small group of people. He was very kind and spoke about life and the mind in a way that profoundly affected her. From that one encounter, Tracy completely changed. She stopped being argumentative with Henry. Instead, she became extremely peaceful and compassionate.

Tracy's transformation terrified Henry. His anger accelerated as he attempted to get Tracy to go back to doing what he had been familiar with his whole life—arguing. As he became angrier, however, she became more resolved. Eventually, Henry saw that whatever she had realized, though it seemed simplistic to him, had undeniably changed her into the loving person she had become. He finally gave in to the feelings that Tracy was now living in and opened his mind to what she had realized, that all of our lives are created moment to moment by thought.

The power of Henry's insight completely changed his life as well. The anger he had carried around all his life vanished, because he realized where it was coming from—his own thoughts. As a result, he began to approach life in a fun-loving way and finally became a happy person after all his years of searching. As the years passed, Henry's insight grew and he went from being a struggling carpenter to a successful business consultant and leadership coach.

I could write several stories about the ways in which Henry has helped businesses to transform through the understanding of the principles of the mind. But the story that most profoundly affected me was the one in which a pivotal insight changed his relationship with his father and how the power of that changed his father as well.

The Fork in the Road

When Henry first began to have insights about his thinking, he realized *his own innocence* in how he had mistreated Tracy and everyone else in his life through his anger. He realized that he had been caught up in a world of his own self-generated thoughts and had acted according to those thoughts. However, he saw that at the core of his being, he was absolutely healthy and beautiful.

One night as he lay in bed just before dropping off to sleep, memories of his father began to flood through his mind. Prior to that night, all he could remember of his childhood were the negative experiences—the abuse, the drinking, the anger. But this night was different, and he was hit with a barrage of positive memories about growing up. To his surprise, he remembered a lot of good in his childhood. He remembered that his father actually did love him and his brothers and sisters and that there were many very good experiences.

At that moment, his whole view of his father changed. In seeing his own innocence, he was also able to see the innocence of his father. In a sense he forgave his father, but in another sense, as he told me, "There was nothing really to forgive. Once I saw his innocence in acting on his thinking, just like mine all those years, my anger disappeared and my love for him took its place. For years I had unsuccessfully tried to forgive him. But when I saw how I was creating my life through thought, I saw how everyone else was doing the same thing. I saw life from a new vantage point."

After that night, Henry called his father for the first time in several years. From that point on he kept in contact with him, though his father was still drinking heavily and none of his other siblings would have anything to do with him, even after his wife died of cancer. Many years passed, and though they talked on the phone, Henry never saw his father, nor did his father meet Henry and Tracy's children. His father's alcoholism continued to progress until one day when Henry was giving a talk, he got an urgent message that his father was dying and only had a few hours to live. Immediately, Henry left the conference and flew several thousand miles to visit his father, who was in the hospital.

On the plane to see his father, all Henry could think about was what he would say to his father to forgive him for all he had done and to ask for his forgiveness for all he had done and said in return. His mind was racing with the images of finally seeing his father after many years and his impending death.

After driving through a snowstorm, Henry arrived at the hospital to what looked like a scene out of a movie. All of his relatives were there, probably thirty people in and around his father's room. His father was somewhere within the press of people, behind the various tubes attached to him and the monitors checking his progress, while nurses and doctors scurried about. Everyone in the family was full of emotions—there was anger, sadness, regret, and yelling and crying—a lifetime of unresolved feelings from an alcoholic's family.

Henry really wanted to talk to his father alone, and his family agreed to give them a few minutes of privacy. As he walked into the room, he expected to see a living version of his memory of his big, tough, construction-worker dad. What he saw instead was a shriveled up, ninety-pound old man who was hours from death. His father's watery blue eyes looked toward Henry, vacant from the drugs, pain, and impending death. Instead of seeing his father, Henry saw another suffering human being dying of cancer, pneumonia, and heart disease. His heart filled with compassion and the script he had prepared in his mind vanished.

Out of the blue, Henry heard himself say, "Do you want to get out of here?"

His father looked up at him and became alert, but looked puzzled.

Henry repeated, "Do you want to get out of here?" His father looked at him and nodded a yes.

This was the first time Henry had eye-to-eye contact with his father. He said, "Look Dad, when you sleep, sleep as deeply as you possibly can. When you wake up, wake up as much as you possibly can. And there's one other thing. You know all this guilt you've got about how you've hurt mom and all of us. You don't have a chance unless you let it go and realize you didn't know any better and you did the best you could. You can't survive unless you let it go."

To Henry's surprise, his father looked up alertly and nodded again.

"There's one other thing," Henry said. "Do you remember when you left Maine as a nineteen-year-old young man and you were full of piss and vinegar and had the world by the tail?"

His father got a glimmer in his eyes and the corners of his mouth formed a grin as Henry continued, "You've got to find *that* feeling again, or you're not going to get out of here."

Tears came out of the corners of his father's eyes, and with that Henry knew he understood. His father passed out into a deep sleep, and Henry walked out of the room, wondering why he had just said what he said. *How could I have made this promise to him?* He wondered.

After he left the room, Henry called Tracy and asked her to Fed Ex to him a video of the family and the house to show to his father. He arranged with the hospital for a big-screen TV to be placed at the foot of his father's bed. When his father woke up the next day, Henry asked him if he wanted to see a video of his grandkids. His father agreed and, though the doctors protested, he insisted on watching the entire video. He started to lighten up and motioned for Henry to come over to him. Then he whispered, "Put the f—ing hockey game on!" Everyone in the room started laughing. After the game he went into a deep sleep. This was his turning point.

Henry's father lived another seven years after that day. He remarried, never took another drink, and reunited with all of his family. For the first time in his sixty-three years, he lived a happy life. The day before he died, he called Henry and thanked him for everything he had done.

The Principle in Practice

When Henry realized how he had innocently created his own life—in particular his anger—through his thinking, he could see how every human being was doing the same thing. When he saw this principle, forgiveness for himself and his father was a by-product of that insight.

Forgiveness is forgetting our old way of viewing the past and seeing it from a new perspective. This new vantage point is a place of understanding and compassion. With this new perspective we can, as the saying goes, "Forgive, but not forget." *True forgiveness is forgetting our old view of the past and remembering it through a higher level of understanding.*

When Henry's father forgave himself for a life of alcoholism and abuse, his innate well-being, both mental and physical, emerged.

Forgiveness is powerful medicine for the body and the soul.

32

DISCOVERING THE JOY OF LOSING CONTROL

Melanie had always been in control. She learned early on that being the one in control was the way to feel good about herself. As a child, she was always taking care of her five younger brothers, tending to their needs and keeping them on a tight rein. As an adult, she became known among her friends as "Dear Abby." No matter how unstable her own life was, she always had a word of advice for others. She genuinely cared about people and tried to help them, albeit in a controlling manner.

With her own six children Melanie believed that "mother knows best." Consequently, she was always trying to tell them "the right way" to live and "the right way" to do things. Somehow she came to believe that her truth was better than theirs, and if they didn't see things her way, well then they just didn't see things clearly. Much to Melanie's dismay, however, her kids didn't always appreciate her advice, no matter how much she believed it was for their own good.

Although Melanie appeared to be very much in control on the outside, inside it was quite a different matter. Her mind was filled with thoughts of worry, judgment, fear, and analysis; she was always thinking about someone's problem and a potential solution.

Not surprisingly, Melanie became a counselor. After all, she had a lifetime of training in giving others advice, being caring, and trying to get others to change—in her mind, the perfect combination of skills for a counselor! In her chosen field, however, which was with chemically dependent people and their families, she saw her clients as resistant and tough to work with. She often felt stressed, mostly because her clients didn't heed her good advice.

Melanie had worked in this field for about twenty years when she took a new job in a treatment program that was based on the principles of mental well-being. She had heard of this psychology before, but didn't really know what it was all about. However, she was very open, read all the books about it, and was trained in this approach.

The Fork in the Road

Melanie shared with me the effects that this training had on her life.

"I started to change very gradually," she said. "I had lots of little insights that changed the way I saw life. It began with seeing my own innocence in all the things I had done in the past, the things that weren't so helpful to others. Instead of feeling guilty about it, I began to see that I was just acting on the thoughts that were in my mind, just like everyone else does. Given the way I saw life at that time, I was doing the best I could.

"I didn't try to change my habit of controlling others, although I certainly was aware that it was a harmful habit. Instead, I gradually noticed it naturally fall away as my understanding grew. As I forgave and accepted myself, I became accepting and forgiving toward others. I started to see that they were doing the best they could, and that they had the answers within them, just as I did. For the first time in my life, *I started to see people as having their own answers.*"

One of the instances in which Melanie noticed this gradual change in herself occurred one morning at home. It began when she asked her husband, "Did you hear the weather report, Honey?"

"Yes, it's supposed to be a great day today," he responded confidently. "Ninety-two and humid."

"What? I heard it was supposed to be miserable—really hot and humid," she retorted, ready for a debate.

"I just know what I heard," her husband insisted.

At that moment, Melanie stopped herself. "Oops! Small-stuff alert!" she chuckled. "I think we're falling into our old habit."

Her husband relaxed too. "Yeah, you're right," he said. "We never have seen eye to eye on the weather."

What Melanie realized was that each of them heard the weather differently through their own thoughts and preferences. She loves cooler weather; he loves hotter weather. One is no more right than the other. Something this small used to be grounds for an argument over who was right and who was wrong. Now she sees it just doesn't matter—we all live in a separate reality.

Melanie began to be genuinely interested in her adult children and how they saw things, rather than trying to impose her views and opinions on them. Again, this was a gradual change for her, but the more she changed, the more they started coming to her for support. They sensed that they weren't going to be judged, criticized, or "fixed" anymore, and they loved being around her. Now Melanie sees the wisdom in her children; in the past, she believed she had the monopoly on wisdom and had to get them to see it her way.

Recently, Melanie told me what this realization means to her. "I feel as if an enormous burden has been lifted from me," she said. "I realize it's okay *not* to be in control."

Just when things were going better than ever for Melanie, she faced the biggest challenge of her life. "I was having dinner at my son and daughter-in-law's home. After I finished my favorite key lime pie, I suddenly felt extremely sick. I knew I needed to go to the emergency room immediately. When the doctors took my blood pressure, it was 40/20—I was having a major heart attack. I had to have surgery and spent the next six days unconscious in intensive care. They didn't know if I would make it.

"When I finally woke up, I felt a deep sense of peace. It wasn't the medications; I was still in a lot of pain. But despite my numerous medical complications, I felt completely at peace. It didn't matter if I died, although I really wanted to live. I just felt a level of acceptance that I had never experienced before. *Everything I had ever believed was so important had lost its urgency.* I didn't need to analyze or control anything anymore. Whatever need to control I had left was now completely gone. In hindsight, I believe my level of understanding about Mind went up dramatically as a result of that medical crisis. I realized that there was something bigger than me and my brain running the show. I could let it all go, and it would all be okay. Losing control is the best thing that ever happened to me. I have finally found true joy and peace."

The Principle in Practice

The need to be in control stems from the basic emotion of fear. When we feel fearful, we feel as if we have lost control of others and of circumstances. For some of us, this leads to an attempt to analyze what might be best for others in order to "fix" the situation. This approach assumes two things: first, that other people lack the ability to resolve their own problems; and second, that we know what's best for them. The only dilemma is to get other people to see this and accept our controlling solution. Since people usually resist our omnipotence, this approach can also be stressful for us.

When we don't trust in the wisdom of "not knowing" for ourselves, we certainly aren't going to see the wisdom of it for others. We will cajole, manipulate, pressure, convince, argue—anything to get them to see it our way. If we succeed, we momentarily feel superior, calm, and in control. However, the other people either feel dependent on us to come up with their next solution or they resent us for trying to run their lives.

People like Melanie who use control as a coping mechanism in life usually suffer from stress, which results from a combination of fear, anxiety, worry, and a busy mind. What they gain from feeling needed they lose by feeling the burden of other people's problems. They also unthinkingly disable those around them. The intentions behind this need to "fix" and control others are misguided, but nonetheless very innocent. It looks like a good idea to try to help others, and if a little advice seems like a good idea, why not a lot of it?

What Melanie didn't fully realize before her heart attack is that there is a greater power that is truly in control of her life and everyone else's—the power of Mind. Mind is the life energy of all things, the deeper intelligence that *really* knows what's best for us. All of us have the power to tap into the intelligence of Mind by listening to our common sense. By realizing this power in herself, Melanie saw that each of her kids, her clients, and her husband all have the same connection to this power. Her only job was to point them to the source of their own innate intelligence. That is how she attained the level of deep peace she has today. By seeing this connection to Mind in herself and in those she truly cares for, she is empowering them to trust in their own wisdom. This is the greatest gift she can give to them and to herself.

Let go of control; you never had it anyway.
Instead, discover the joy and peace of knowing
that there is an ultimate power—Mind.

33

HOPE IS THE CURE

Carl has traveled a difficult path during his thirty-two years. A Native American born in Kansas, he was severely beaten and tortured by his biological parents till age three, when he was removed from his home by the state and put up for adoption. By the time the authorities finally intervened, Carl's body was covered with scars from cigar burns, and most of his bones had been broken at one time or another from severe beatings. He remembers none of this early abuse; he knows only what he read in the official reports and the stories he heard from his social workers.

His grandmother was the person who turned his parents in when they resorted to chaining him to a tree with a dog collar like an animal. His relief, however, was short-lived, for he was adopted by a minister and his wife who were just as abusive as his biological parents. As he continued to share his gruesome story with me, I became almost sick to my stomach. Carl spoke of how his adopted mother treated him like a slave, making him clean the house with a brush till the wee hours of the morning. He would be beaten severely while tied to "the rack," a bed in the basement fitted with harnesses to restrain him and the other children. If he bled from the beatings, he would be further beaten for soiling the sheets. He would often pass out, soil his pants, or have convulsions from the extreme pain. To improve his posture, Carl's adoptive mother would force him to stand outside in the winter in his underwear with a dixie cup on his head till it froze.

Throughout his entire childhood, Carl endured these tortures and fell deeper into despair and self-hatred. He believed he must have been a really bad person to be treated this way. During adolescence, he became an alcoholic—an escape from the pain that seemed to work, at least for the moment. Although his parents actually sought therapy for the family, the therapists never believed his stories of abuse; after all, his parents were respected ministers.

The rest of Carl's life has been spent in institutions, mostly for alcoholism. He has tried suicide three times; once he jumped off a bridge with a noose around his neck, only to be saved just before death by people who came by and saw him hanging. Once he was discovered buried in a snow bank; he had been buried there by other skid-row alcoholics. When the snowplow caught his foot and threw him into the street, he was blue, frozen stiff, and seemingly dead. In fact, Carl has been pronounced dead three times from exposure or alcohol poisoning, but, miraculously, he was brought back to life each time. *There must be a reason I am still alive,* he reflected. *I should have been dead scores of times.*

The last time he was in detox, he stayed for three weeks, determined that this time he was going to turn his life around. He was assigned to a treatment center whose approach was based on the three principles. *Maybe this new approach will work for me,* he thought.

The Fork in the Road

He shared his story of transformation with me.

"This program saved my life. I know now that I don't have another drunk in me. The only way I will ever stop drinking is by understanding my thoughts and living in the moment. When I am in this moment, all I have is my thoughts and I know I can change my thoughts at any time, and that changes my reality. I was so caught in the past that my mind was full of painful memories all the time. Now I see that the past is only a thought carried through time. It has set me free. I see now that when I am about to do something stupid like hit someone, I know it is my thinking that is creating that, not the other person, or my past. Now I truly believe I can change.

"I only wish I had learned this years ago. I could have avoided so much of my pain. I am so happy now and love myself from my core inside. I will devote the rest of my life to bringing this to Indian children in the schools by telling them my story and what I have learned."

I asked Carl, "How did this transformation occur?"

"I came with an open mind. I knew this program was different and that I was alive for a reason. They told me about thought and my moods and how to clear my mind. I learned how to get into a healthy flow of thought—to step from the grunge to my serenity. I realized I didn't have to think all the time, thinking, thinking, thinking, a really busy mind. I learned that I could trust a quiet mind—it would take care of me. It is like when I play hockey. I am a goalie and when my head is clear, they can't get the puck past me. When I am processing my thoughts, I can't see the puck. When I saw that connection, I knew my life would change. My life is simple now. I know where my life comes from now—my thinking. This approach is so similar to what my Native elders have taught for centuries. Our elders' elders have been telling stories similar to this philosophy forever.

"My life is beautiful now. I can listen to people now. I can teach this to others because it is so simple. I have so much hope for myself and others. I had thought I would always be a drunken Indian, but I now see that I can be a whole human being. I am alive for a purpose. I owe that to my people. I learned this program in three days. . . . I had been through treatment fifteen times and rehashed my past innumerable times. I knew that wasn't the answer. I was starving for this knowledge. If I could learn it, anyone can."

The Principle in Practice

How can a life of trauma, abuse, alcoholism, despair, and suicide be transformed with such apparent simplicity? Neither Carl's past nor the severity of his problems hindered his transformation. Chemically dependent people, with addictions both less and more severe, could tell similar stories about the recovery of their health.

Carl came to treatment with an *open mind*. That gave him *hope*, and he was in a position to *listen*. He began to understand the principles underlying his psychological functioning, and he had the power to *wake up the health that was already within him*. No matter how much a human being has been abused, physically or emotionally, his or her innate mental health cannot be damaged. Like a dormant seed, all it needs is water and light—love and understanding—to come back to life. The simplicity of the three principles, Mind, Thought, and Consciousness, awakened in Carl the power to transform his life.

Never give up hope. Hope has the power to open your mind, allowing the insights that can change your life to bubble to the surface.

34

"GIVE UP YOUR GOALS SO YOU CAN TRULY SUCCEED"

LAO-TZU, FROM THE *TAO TE CHING*

Not long ago, Walter and Lisa were a young married couple perched on the edge of success. When he completed his master's degree in psychology and business, his goal was to work in management, consult with companies about the human-relations environment, or do executive coaching. When the desired positions did not come immediately, Walter felt he had no choice but to put his goals on hold for a few years. A native of Germany, Walter seemed to be able to get nothing but translation work.

Both he and Lisa dreamed of moving back to his native country. He would be able to be with his family, and she could learn the language and culture. They were adamant about wanting to go to Germany, get a good-paying job, and travel before they had children. In an attempt to make their dream a reality, they spent six weeks in Germany sending out resumes, placing ads in the paper, and putting their determination to work. On the last day before they were to return to the United States empty-handed, Walter got a call for an interview.

"At last our dreams can come true!" Lisa exclaimed.

But Walter was cautious. "Don't get too excited, Lisa. It's only an interview," he said.

After Walter's interview, he and Lisa returned to the United States, expecting a call any day from the company. One week passed, then two weeks, then a month. Lisa impatiently pressured Walter to call them, but he knew he had to wait for them to call him. Finally the personnel director called around Christmas and said they were definitely interested in him, but the decision-making process was slow. After a few more difficult months of negotiation, the dream finally came true. The pay was great, and it was an excellent job that could lead to a position as vice president. The company would also help pay off his school loans and pay for the move. It couldn't have been a more perfect job offer.

Walter and Lisa were ecstatic. They had finally *made it happen.*

Walter went on ahead to Germany to begin his new job, an enviable position by German standards and perfect for him and Lisa. He would be on probation for two months to see if he was right for the job and the job was right for him. In the meantime, Lisa quit her job and started packing for the big move.

The Fork in the Road

From the time he began work, Walter had an uneasy feeling about the company and his job. Something just didn't feel right, but he would dismiss such thoughts as insecurity. He was afraid to even think of going back to the United States after he had burned his bridges, afraid to consider giving up what appeared to be the perfect career move for him. But finally he realized that he could no longer ignore his feelings. One night he went for a long walk alone and reflected on his dilemma. He began to cry as he realized that the job just wasn't right for him. But instead of sadness, his tears were tears of joy and relief at his realization.

Walter called Lisa with trepidation, because he knew she would be disappointed and maybe even angry. He said to her, "Lisa, I don't want to stay in the job. I don't like it. It doesn't feel right to me. I know I won't be happy here." Lisa waited during the long pause on the other end of the line.

Lisa could tell from the strength of conviction in Walter's voice that he was sure. She also knew that as disappointed as she was to give up her dream, she needed to support Walter in his happiness. Because of the clarity of her husband's feelings, Lisa decided to trust in the unknown that was to come.

The next day Walter went into his CEO's office and handed in his resignation. The CEO was shocked and disappointed, but knew he had no choice but to accept it. "Why are you leaving and what will you do now?" he asked, genuinely puzzled.

Walter proceeded to recount to the CEO a series of problems he saw in the company and explained that what he found there was quite different from the way the job had been presented to him. Walter then told the CEO that he decided to start his own consulting company. He shared this with such clarity and confidence that the CEO was very impressed and asked, "What will you do for companies as a consultant?"

Walter pondered for a moment, since he hadn't fully formulated his vision yet. Then he just spoke from the heart for quite a long period of time. He told him how thought creates our reality and how each human being has an innate capacity for motivation, common sense, and other qualities needed in an effective employee. The CEO was so impressed that he asked Walter if he would be a consultant to him and the company. Suddenly Walter had his first consulting client!

Before he left Germany, Walter did a one-day training seminar with the CEO and made plans to train more of the company managers in the principles of leadership and mental well-being. To date, he has been to thirteen countries and trained over two thousand employees. It has gone so well that he is doing more within the company that originally hired him and has other client companies in Germany and in the United States.

What looked logically like a foolish move in quitting his job turned out to be better than anything Walter and Lisa could have imagined—he made more money as a consultant, he had total freedom, and he could have the best of both worlds by commuting between the United States and Germany. Lisa was able to live in Germany for several months and traveled with Walter as his consulting business took him all over the world.

When Walter and Lisa shared their story with me, Walter said, "Our goals were so tiny compared to the dreams our hearts had in store for us. We never would have imagined we would be doing what we are doing now. We have a new home, a good income, a great career, and most importantly we are happy. By following the 'heart track' rather than the 'career track,' we are far better off."

Lisa added, "Instead of ignoring our feelings of the heart, we listened and trusted them, even though the future was unknown. If we had gone the other way out of fear, we would have been so limited."

The Principle in Practice

We all formulate goals for ourselves based on what we think will make us happy. If these goals are based on conditioning and habit learned from our past, then even if we attain those goals they will feel empty after a while. As a result, we set a new set of goals, hoping these too will make us happy.

On the other hand, if we are willing to give up our habitual goals and truly listen to our hearts, as Walter and Lisa did, we can discover something much greater and more deeply rooted in the core of our being. When we follow the *path of the heart*, we are listening to the deeper intelligence of wisdom. Trusting the unknown feeling of the heart can initially take courage, until we discover through *the leap of faith* that it always works out beyond our wildest imagination.

As we increasingly trust in our deeper feelings about decisions we face in life—whether they are about taking a job, choosing a mate, making a move, or going back to school—our faith will be transformed into knowing. Walter felt very serene after that night when he took the walk and cried. He *knew* what to do. When we trust this *knowing*, our feelings become peaceful. Before Walter acknowledged his feelings, however, he felt distress. Feelings are the best compass to help us navigate into the unknown of the future.

> **Trust in the invisible force of feelings. They know more than your brain could ever know about making your dreams come true.**

35

A COMMUNITY LEARNS HOW TO BE CALM, COOL, AND COLLECTED

This story is a little different from the others in this book. It is the story of how a whole community discovered the health that is within it.

The people of Mount Hope, a neighborhood in the South Bronx of New York City, had lost all hope. The hallmarks of the neighborhood were crime, drug dealing, deteriorating housing, homelessness, a lack of opportunities, and a feeling of isolation. The people were fast-moving, irritable, angry, reactive, loud, and full of intensity—what some people would call typical New Yorkers.

Roberta had grown up in Mount Hope and had raised her family there. As a black woman, she wanted to succeed in corporate America and worked for thirty years as a sales manager for a major corporation. She was stressed out and competitive, and she was so preoccupied with her work that she had become oblivious to the deterioration of the neighborhood around her.

One day Roberta woke up to what was happening in Mount Hope and decided to do something about it. She formed a grassroots neighborhood revitalization group to save the area. The group made some progress in getting the drugs out of the community and improving the housing. And as she grew weary of her corporate job and a business climate of downsizing and fear, she saw this community group as her way out. She decided to devote herself full-time to administering the neighborhood housing agency, which by now was a multimillion-dollar corporation in need of her managerial skills. To Roberta, this new job sounded like it would be a piece of cake compared to the dog-eat-dog corporate world she had come to know. She was in for a big surprise.

When she told me about the conditions she encountered at the neighborhood housing agency, she said, "It was ugly. Nobody talked to anyone else. There was fighting among those who were supposed to be a team, there was no respect between staff members, and people were rude. The corporate world looked friendly compared to this group of people, and all I could think was, what have I gotten myself into?"

One of the advantages of her new position, however, was that their funding agency was very supportive of their needs. They recommended that the entire staff be trained in the three principles of mental well-being. It had worked in other similar neighborhoods with success. The focus was to first help them find their own healthy state of mind, so that they could help others.

The Fork in the Road

Roberta was very hopeful about the new consultant's message, but wondered how this soft-spoken white male could ever reach a group of intense, stressed out, loud black and Latino human services workers from New York. The first morning of the training, they began in their usual loud, combative, and restless style. By the end of the morning, however, there was a *stillness* in the room that shocked Roberta.

Ned, the consultant, had become a pioneer in the field of resiliency and prevention of social and psychological problems. The philosophy was to work with the mental health of the community, rather than to impose programs on the community from the outside. Once people in communities found their mental bearings, they were capable of coming up with their own solutions to solve problems.

Roberta said to me, "I knew we had found the missing link in becoming effective agents of change. We had good people, money, and resources, but we didn't have healthy minds. This could be the answer we had been seeking without even knowing it. I also realized that this was not only for them, but that I needed it as well. I had become more stressed than I had ever been in the corporate world. The night I came home from the seminar I was determined to apply what I had learned to my personal life. I immediately realized how poorly I had been treating my husband. Most nights I would barely talk to him, and if I did, I was angry and judgmental. I was so 'out of it' from all the stress.

"Jon, my husband, was a hardworking, kind, and supportive mate. Every day he would make our bed without me even asking. Every night I would return home from work and immediately tear the bedspread off, because he hadn't done it properly. *Didn't he know how to make a bed correctly?* I would often wonder. *The rounded part is supposed to go at the bottom, not the top!* I would fume around the house all night about this; he had no idea why. The night of the seminar, I came home peaceful and loving, and I didn't tear the bed apart. Instead I realized how lucky I was that he even made the bed, and I thought about all the other things he did. I was loving toward him, and I could tell he didn't know what was up.

"After a pleasant evening, we sat down and I asked him if he knew what was wrong with the way he was making the bed. After telling him the proper way, he said, 'I had no idea. Why didn't you tell me? Thanks, I'll do it differently from now on.' I expressed my gratitude for just how kind he had been. Since then, my whole life has changed. I became calmer, more patient, more accepting, and I saw health in all people, even the drug dealers. I realized my New York pushiness wasn't necessary to get the job done. My marriage and the whole agency have changed."

After a while, the human services staff at the housing agency began to train the residents of the housing project in the three principles. Prior to the training, the residents were much as the staff had been—divided, hostile, uncooperative, and territorial. In addition, they only had 55 percent occupancy, and of those only 49 percent paid their rent. The children destroyed the buildings with graffiti, trash, and vandalism.

After training the residents, the neighborhood began to shine. Occupancy in the housing project went up to 90 percent, and rent collection went up to 92 percent. The graffiti disappeared, and people took pride in how the neighborhood looked.

Repairs and accompanying costs went down dramatically. The staff and the residents started to experience what mental well-being feels like—they were calm, patient, loving, happy, and thoughtful of others. With these feelings, they naturally became proud of their neighborhood and did what they could to improve it. Many of these people had been homeless and mentally ill before they turned their lives around. Joline, for example, was released from the state hospital when it was closed down in the 1970s. She had spent the last twenty years living in alleys and on the street. After attending a series of classes on mental health she is living in public housing, going back to school, and has lots of friends. Most important, she is hopeful about her future and the possibility of being "normal." She is typical of many residents. Finally the residents of Mount Hope were living up to their name.

Health became the new norm for Mount Hope residents. Anger used to be the only way to get anyone's attention or to get anything accomplished. Now anger became a signal that they were off their center, and when they saw it, they would remind each other to stay "calm, cool, and collected." That became their motto. Once they understood the role of thought in creating their experience and their lives, they were truly empowered to change themselves and their neighborhood.

The Principle in Practice

In our efforts to fight crime, poverty, homelessness, and neighborhood blight, we have continuously thrown money and programs at our poor, inner-city neighborhoods, but the results have been temporary at best. That is because the root of any problem, like the root of any weed, must be removed if the problem is to be eliminated, thus allowing the green grass and flowers to grow freely.

In Mount Hope, Roberta and her co-workers discovered how to get to the root of the neighborhood's problems and thus create lasting change. The root of all human behavior is *thought*. When the staff and residents learned how their thinking created their experience and how to access wiser thought, they were able to eliminate their unhealthy ways of thinking. They learned to recognize the feelings and emotions associated with their thinking and thus tell the difference between healthy and unhealthy thoughts. In so doing, they changed the way they dealt with each other and themselves. They started to see possibilities and create solutions, instead of living within the wall of hopelessness that had previously surrounded their thinking. And now, hope springs eternal in Mount Hope.

Change within a community begins within the thinking of each of its members. This is the way to true empowerment.

36

CHANGING THROUGH SELF-ACCEPTANCE

Kara never really enjoyed dating or being in relationships. From the time she was an adolescent until recently in her forties, she has always seen dating and relationships as stressful and more of a chore than a pleasure. She didn't understand how anyone could enjoy them, because they seemed to be fraught with feelings of insecurity, expectations, disappointment, and heartache.

When Kara was attracted to a man who was interested in her, her habitual pattern was to begin to have expectations. Her mind would fill with thoughts like, *Will he call me this weekend for a date? Why hasn't he called yet? He must not like me as much as I like him. Why does this always happen to me? Men are all jerks! Maybe there's something wrong with me.* Kara would go from one expectation to another and jump to conclusions about any unmet expectation. As a result, she doubted herself and felt resentful, hurt, or disappointed.

To cope with this vicious cycle, Kara would vacillate between avoiding dating all together and trying to think positively. She would try to figure out why she didn't have a relationship: *Am I not physically attractive enough? Am I too independent and self-confident and thus threatening to men?* The bottom line was that Kara was spending a great deal of time thinking about relationships and never coming up with any new thoughts about the subject. No wonder she didn't enjoy dating!

A few years ago, Kara began to learn about the principles of healthy thinking, and her life changed in many ways—she became more confident in her job, and her family relationships improved greatly. Yet her insecurity about dating persisted. Dating was truly her blind spot.

However, several insights began to pave the way for a breakthrough in how she approached dating. The key for Kara was the insight that instead of trying to change herself by analyzing her problems, instead of judging herself and trying to fix herself by getting all kinds of advice, *she accepted herself as she was.* She discovered a paradox: When she accepted a particular characteristic about herself, she would change almost automatically. She would say things to herself when she saw an old unhealthy habit like, *Oh well, here I am again. Maybe next time I'll do it differently.* In short, she wouldn't make a big deal about it and that would free her to have insights into her behavior.

The Fork in the Road

Kara started to accept the fact that she was not in a relationship. As she accepted this fact without any judgment, she gained a certain amount of humility. She realized that she knew very little about relationships or how to go about having one, but instead of feeling bad about this, she became a student of relationships. As she told me, "If my relationship blind spot was a twenty-five-ton rock in the middle of the road, self-acceptance was the stick of dynamite that blew the rock to pieces. The second step was to become a student of relationships."

Kara asked her friends who were good at relationships questions like, "How do you ask someone out? How do you handle it when they say no?" To Kara it was a huge deal to ask a man for a date. It carried the same level of anxiety as asking someone to marry her. But when she asked her friend for advice, her friend said to just call a man, have a chat with him, and then see if it felt right to ask him out. Kara thought, *I can have a chat with someone easily. That's no big deal.* She had never considered that she could just have a conversation with a man. Approaching it in this way made the prospect of dating begin to sound easy and fun.

Kara began to ask men out, take risks, and then reflect on what happened. Before, she would have discounted her friends' advice or she would have used it as ammunition to judge herself. But now she had nothing to lose, because she let go of her expectations. She was looking at her dating experiences as opportunities for learning and having fun. If her old insecure thinking did crop up, she would recognize it and be interested and amused by it. She also recognized how the insecure thinking of the men she dated was similar to hers, and instead of interpreting it negatively, she would feel compassion for them. The whole dating scene has become a process of discovery for her.

The Principle in Practice

Bookstores are filled with self-help books on how to find a date, become more physically attractive, or learn foolproof lines to entice a man or a woman into going out with you. Through the principles of healthy thinking, Kara discovered a different approach to change, one that begins with self-acceptance. When we accept ourselves as we are in all our imperfections, we can also accept that we don't know much about a particular subject. When we feel acceptance of ourselves in that state of not knowing, we naturally become curious and start to learn. However, until we accept ourselves as we are now, we turn any new information into a weapon of self-judgment.

When Kara accepted the fact that she had no clue as to how to have a relationship, she started listening to her friends' advice and taking it to heart. She didn't blindly accept their advice because everyone is different, but she did broaden her own view of things. When we see our blind spots as areas of discovery instead of reasons for shame and embarrassment, we learn and gain insights into ourselves.

When we recognize our negative thinking without judging it, it can come out into the light where we can see it more clearly and *admit what we don't know*. Not knowing is the threshold of insight. By admitting we don't know with acceptance rather than judgment, we put ourselves at the "bus stop" of insight. If we aren't at the bus stop, we won't catch the bus no matter how often it comes. Acceptance is what puts us at the bus stop.

Change is easy when you accept yourself fully as you are now, instead of trying to change yourself through effort and judgment.

37

THE POWER OF THE HEALING MOMENT

Bobby was hurt during a high-school hockey game. He knew the instant he was hit that he was paralyzed. He broke his neck and two cervical bones. The prognosis was grim.

Bobby's mother, Natalie, could never forget the doctor's words. "The prognosis," he said, "is not good. I am sorry, but this is a very, very serious injury."

So this is it, Natalie thought, numb with shock. *You wonder what it is going to be like to face your darkest moment and this is it. How am I going to deal with this?*

The Fork in the Road

Natalie shared how she transformed her despair into hope.

"As our son was in surgery and our friends came, my mood began to lift," Natalie said. "A lot of what my husband and I had learned about the principles of mental well-being began to settle in, and I knew to stay in the moment. Whenever we would begin to jump into the future, we would become terrified. We knew not to go there instinctively. The same was true of going back into the past. We would get so sad if we did that. We had all we could handle in the present.

"Then it began to be clear that, in the moment, all possibilities exist. No one knows what the future holds and no reality is absolute, including the reality that Bobby's doctors presented to us. I saw that anything the doctors said, in their innocence and love, was the best they could tell us based on their past information and experience. And that is all I knew it to be. Their prognosis was very grim. All we knew was that we had all we could deal with in the moment. I began to remember all that I had learned about thought. I knew that if I saw this event with my son as tragic, that is what it would be for us and him. I also saw that if I knew he would be all right, no matter what the outcome, he would be all right.

"Bobby spent fourteen hours in surgery the first two days. Between surgeries, the media interviewed us. I was amazed at the outpouring of support from our community, the prayer chains. Over six hundred people came to church to pray for Bobby.

"I told the news people that we knew that Bobby would be all right. The essence of Bobby could not be damaged. In my wildest dreams I wouldn't have believed that we would have the wherewithal to deal with the media at such a time. It became so apparent that if we stayed in the moment, that was all we needed. There was the gift of understanding, and there was this outpouring of love and support from our friends and our community. We knew we didn't have to worry about anything. Everyone became the face of God. It was a profound feeling.

"The hospital said they had never seen so many visitors for any patient. A friend said that people were drawn to the positive feeling of total peace, love, and understanding. People who worked at the hospital would just come around, drawn to that magnet of positive energy.

"We knew that we didn't want any negative feelings to enter Bobby's room. We wanted to surround him and ourselves with only loving and positive energy. And if we stayed in the moment, it was always there.

"One of the most profound moments I can remember with Bobby was when he first moved his thumb and index finger. This meant he could use a joystick. The doctors were still pessimistic about the future, but we felt grateful for any positive sign. As the doctors left, Bobby said, 'God knows more than they do. All I want to do is be happy no matter what.' I reassured him confidently that he could definitely be happy, no matter what the outcome. That feeling was inside of him. That was all that mattered to Bobby.

"Almost instantly, Bobby was filled with positive feelings. The more he noticed how good he felt, the better he felt. He was overcome with how good he felt for three days. He stayed up till late at night talking and being so hopeful."

It was then that Natalie knew he would be okay.

"From then on I would remind him that if he had that good feeling at his darkest hour, he could have that feeling any time, unconditionally."

I asked Natalie if she had any low periods during this time.

"Sure I did," she said, "but in my low moments I knew it was just thought. My counselor would tell me to think of my thoughts like raindrops on the windshield. If I just kept the wipers on, they would go away. I knew I couldn't afford any negativity or I would plummet. When others would come in the room and begin to commiserate about how awful it was, I would tell them about how it was all thought, and my mood would lift and so would theirs. I began to see that when I shared this understanding, it lifted me up. It was a constant reminder to me of how we all create our own reality and perception.

"I could even see that the doctors' sense of urgency and impatience was just their low mood, and when I reminded myself of this, it protected me from getting caught up in their urgency. Their fear of the insurance running out before Bobby was well was just their fear.

"So many times my wisdom would come through for me. Whether it was bringing in soothing baroque music or not having visitors at that moment. I saw that my wisdom would give me just the right thought or insight at just the right moment. My analytical thinking could never do that to the degree my wisdom did. I knew that if I trusted in my innate mental health, I would be okay. All of the members of my family were constantly reminding each other that if we got in a low mood, all we needed to do was trust in our wisdom."

The wisdom of living in the moment guided Bobby, Natalie, and their whole family throughout that very trying time, as it still does in Bobby's continuing rehabilitation and additional surgeries. I recently heard that at Bobby's high-school graduation ceremony, he walked down the aisle to receive his diploma. It appears that Bobby was right; God knows better than the doctors. The important point was not whether Bobby ever walked again, but the grace and peace with which this family went through this trying time together—in the moment, one moment at a time.

The Principle in Practice

I don't believe I have ever heard a more powerful testimonial on the power of living in the moment. Through her understanding of thought as the source of her experience, Natalie realized that she had a choice as to how she responded to Bobby's injuries, and there really was no choice but to stay in the moment. She knew that only in the moment would she be protected by loving feelings and have all the insights she needed. It was only in the moment that Bobby's body would heal at its best, that they would be guided to the next step, and that they would be protected from other people's lack of understanding and even their own low moods. It was only in the moment that they would not fear. Instead of fear, they would have faith and peace of mind.

Not only were Natalie, Bobby, and their family helped by an understanding of the power of Mind, Thought, and Consciousness, but a whole community was deeply touched and brought together by the power of love that is at the heart of true understanding.

No matter what challenges you face, trust in the power of living in the moment for healing, living, and loving.

38

YOUR OWN STORY

Now that you have read *The Speed Trap*, you may have had some changes in your own life. Write them down for yourself. Telling your own story of transformation, no matter how small, can help you see the power you have to create your life. If you care to share it, you can send it to me.

RESOURCES

Health Realization (formerly known as Psychology of Mind) has been applied to a variety of topics and populations, including: counseling and psychotherapy, substance abuse, mental health, business consultation, primary prevention, community revitalization and empowerment, medicine and psychiatry, health care, corrections and policing, and numerous other areas. For more information, or to send me your story, you may contact me at my web page at:

www.thespeedtrap.com

or

Joseph Bailey
P.O. Box 25711
Woodbury, MN 55125-9998

Health Realization Materials

For books, tapes, videos, articles, newsletters, and information on Health Realization, contact:

The Pyschology of Mind/Health Realization Resource Center
Website: www.pomhr.com
Phone: 800-481-7639
Fax: 541-383-5149

Training Programs

Phone: 800-781-2066